ANALOG INTERFACES FOR DIGITAL SIGNAL PROCESSING SYSTEMS

THE KLUWER INTERNATIONAL SERIES IN ENGINEERING AND COMPUTER SCIENCE

ANALOG CIRCUITS AND SIGNAL PROCESSING

Consulting Editor

Mohammed Ismail
Ohio State University

Related titles:

SYMBOLIC ANALYSIS OF ANALOG CIRCUITS: Techniques and Applications, Lawrence P. Huelsman, Georges G. E. Gielen
> ISBN: 0-7923-9324-4

DESIGN OF LOW-VOLTAGE BIPOLAR OPERATIONAL AMPLIFIERS, M. Jeroen Fonderie, Johan H. Huijsing
> ISBN: 0-7923-9317-1

STATISTICAL MODELING FOR COMPUTER-AIDED DESIGN OF MOS VLSI CIRCUITS, Christopher Michael, Mohammed Ismail
> ISBN: 0-7923-9299-X

SELECTIVE LINEAR-PHASE SWITCHED-CAPACITOR AND DIGITAL FILTERS, Hussein Baher
> ISBN: 0-7923-9298-1

ANALOG CMOS FILTERS FOR VERY HIGH FREQUENCIES, Bram Nauta
> ISBN: 0-7923-9272-8

ANALOG VLSI NEURAL NETWORKS, Yoshiyasu Takefuji
> ISBN: 0-7923-9273-6

ANALOG VLSI IMPLEMENTATION OF NEURAL NETWORKS, Carver A. Mead, Mohammed Ismail
> ISBN: 0-7923-9040-7

AN INTRODUCTION TO ANALOG VLSI DESIGN AUTOMATION, Mohammed Ismail, Jose Franca
> ISBN: 0-7923-9071-7

INTRODUCTION TO THE DESIGN OF TRANSCONDUCTOR-CAPACITOR FILTERS, Jaime Kardontchik
> ISBN: 0-7923-9195-0

VLSI DESIGN OF NEURAL NETWORKS, Ulrich Ramacher, Ulrich Ruckert
> ISBN: 0-7923-9127-6

LOW-NOISE WIDE-BAND AMPLIFIERS IN BIPOLAR AND CMOS TECHNOLOGIES, Z. Y. Chang, Willy Sansen
> ISBN: 0-7923-9096-2

ANALOG INTEGRATED CIRCUITS FOR COMMUNICATIONS: Principles, Simulation and Design, Donald O. Pederson, Kartikeya Mayaram
> ISBN: 0-7923-9089-X

SYMBOLIC ANALYSIS FOR AUTOMATED DESIGN OF ANALOG INTEGRATED CIRCUITS, Georges Gielen, Willy Sansen
> ISBN: 0-7923-9161-6

STEADY-STATE METHODS FOR SIMULATING ANALOG AND MICROWAVE CIRCUITS, Kenneth S. Kundert, Jacob White, Alberto Sangiovanni-Vincentelli
> ISBN: 0-7923-9069-5

MIXED-MODE SIMULATION: Algorithms and Implementation, Reseve A. Saleh, A. Richard Newton
> ISBN: 0-7923-9107-1

ANALOG INTERFACES FOR DIGITAL SIGNAL PROCESSING SYSTEMS

by

Frank Op 't Eynde
MIETEC Alcatel

Willy Sansen
Katholieke Universiteit Leuven

KLUWER ACADEMIC PUBLISHERS
Boston / Dordrecht / London

Distributors for North America:
Kluwer Academic Publishers
101 Philip Drive
Assinippi Park
Norwell, Massachusetts 02061 USA
617-871-6300
Distributors for all other countries:
Kluwer Academic Publishers Group
Distribution Centre
Post Office Box 322
3300 AH Dordrecht, THE NETHERLANDS

Library of Congress Cataloging-in-Publication Data

Eynde, Frank Op 't, 1963-
    Analog interfaces for digital signal processing systems / by Frank
Op 't Eynde, Willy Sansen.
    p. cm. -- (The Kluwer international series in engineering and
computer science. Analog circuits and signal processing)
    Includes bibliographical references and index.
    ISBN 0-7923-9348-1 (alk. paper)
    1. Linear integrated circuits. 2. Digital integrated circuits.
3. Signal processing--Digital techniques. 4. Metal oxide
semiconductors, Complementary. I. Sansen, Willy M. C. II. Title.
III. Series.
TK7874.E96 1993
621.39'814--dc20   93-17416
                                          CIP

Copyright © 1993 by Kluwer Academic Publishers

Printed on acid-free paper.

Printed in the United States of America

TABLE OF CONTENTS

ANALOG INTERFACES FOR DIGITAL SIGNAL PROCESSING SYSTEMS

FOREWORD

It is a great honor to provide an introduction for Dr. Frank Op 't Eynde's and Dr. Willy Sansen's book "Analog Interfaces for Digital Signal Processing Systems".

The field of analog integrated circuit design is undergoing rapid evolution. The pervasiveness of digital processing has considerably modified the micro-system architectures: the analog part of complex mixed systems is more and more pushed at the boundary limits of the processing chain. Moreover, the increased performance of digital circuits, in terms of accuracy and speed, are making the specification requirements of analog circuits very strict. In addition to this, the technology, supply voltage and power consumption of analog circuits must be compatible with those, typical for digital circuits. Therefore, in a few words, analog circuits are becoming complex and specialised interfaces between the real world and digital signal processing domains.

This technological evolution should be accompanied by an equivalently fast evolution in designer competencies. Knowledge of complicated signal handling should be quickly replaced by know-how of simple but very accurate and very fast signal processing and a solid background in data conversion techniques. All of this through the use of the CMOS (and possibly BiCMOS) technology.

Obviously, a new approach is needed in the design methodology; an approach that, on one hand must return the transistor-level requirements to achieve advanced performances, and on the other hand focuses the designer only on the few methods that are suitable for mixed-mode technology implementations. This new trend is a reality that can not be overlooked or ignored. It is, therefore, gratifying to acknowledge the work done for this publication. It is an answer to the new needs of the analog designer's community.

Franco Maloberti
Professor of Microelectronics
University of Pavia, Italy

PREFACE

Already in the earliest days of the electronics era, somewhere around the beginning of this century, an important part of electronic equipment was intended to perform operations on analog electrical signals, generated by for instance a telephone set, a microphone, a television camera or other kinds of electrical sensor. Examples of such signal operations are amplification, filtering, addition of two signals, storage in an analog memory, and nonlinear compression or expansion.

Common to all these signal processing systems is the requirement to preserve the signal information: during the various signal processing steps, the signal information should not be corrupted by noise, by spurious signals or by undesired nonlinear effects. The high-frequency signal contents should not be distorted by bandwidth limiting effects. This requires electronic circuits with a sufficient bandwidth and dynamic range. Prior to the mid 1970s, all signal processing was performed by analog circuits, suffering from component noise and component nonlinearities. Developing such signal processing system for high-performance applications with a high dynamic range was therefore a costly, time consuming task, requiring highly-skilled circuit designers.

Since the mid 1970s, digital circuits with an ever increasing number of functions and an increasing speed performance became available at a continuously reducing cost. These circuits offer an alternative for the classical analog signal processing.

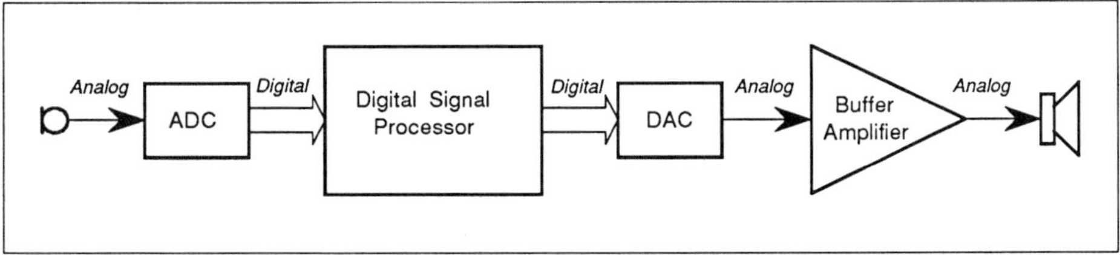

Fig. 1: The basic contents of a digital signal processing system

Fig. 1. shows the basic contents of a digital signal processing system. After converting the physical input quantity to be measured (e.g. air pressure variations in an audio system) to an analog input signal with a proper sensor (e.g. a microphone), the analog signal is converted to a digital format with an **Analog-to-Digital converter** and applied to a Digital Signal Processor (DSP). In this processor, digital mathematical signal operations such as amplification, addition, digital filtering or digital nonlinear expansion or compression are performed. The output signal is re-converted to analog with a **Digital-to-Analog converter**. An analog **Buffer amplifier** drives an actuator (e.g. an audio speaker) which generates the physical application output quantity. This approach has found wide spread, for instance in audio equipment: where a classical audio system was fully analog, modern recording techniques convert the analog signal to a digital format. A similar technique is used in modern telecommunication systems: where the classical telephone set was fully analog, modern ISDN networks and cellular radio systems are based upon digital data transmission. High-definition television and automotive electronics are other applications of digital signal processing.

By increasing the word length and the clock rate, the signal degradation in the digital circuits can be made very small. Therefore, complex signal operations can be performed with an accuracy that is unfeasible with classical analog signal processing systems. As a result of this technological breakthrough, customers have increased their signal processing demands over the years in all the sectors of the electronics industry. For instance, where the classical HI-FI norm required a dynamic range of 60 dB, modern digital audio equipment achieves a dynamic range of at least 100 dB. And where a classical telephone set achieves a dynamic range in the order of 50 dB over a voice band of 3 kHz, a modern ISDN link requires signal processing components with a dynamic range of 72 dB over a 70 kHz bandwidth.

With these increased demands, the analog interface circuits - the ADC, the DAC and the output buffer in Fig. 1 - become the bottle necks in the signal processing chain. In order to fully benefit from the speed, the accuracy and the robustness offered by digital circuits, fast analog interfaces with a high dynamic range are required.

This is a new challenge for the modern analog circuit designer, requiring new design approaches and novel circuit design techniques. In this book, the practical implementation of the three classes of analog interfaces - ADC, DAC and buffer amplifier - is studied.

This book is organised as follows:

Operational amplifiers are the key building blocks in most high-performance analog systems. In the chapters 2 to 5, the system performance of various interface circuits is expressed in terms of the amplifier characteristics such as GBW, DC gain or CMRR. It is shown there that a high dynamic range often requires amplifiers with a large GBW. In an introductory **Chapter 1**, the possibilities and limitations of the CMOS technology for the implementation of wideband amplifiers are investigated. Fundamental relations are derived between the amplifier GBW and its power consumption. The capabilities of the CMOS technology are illustrated with two design examples.

In **Chapter 2**, the design of Low-distortion power amplifiers is studied. The importance of second-order effects such as the nonlinear Common-mode gain, the nonlinear Power-supply gain or the nonlinear thermal feedback are illustrated. Ultimately, the distortion is determined by these effects rather than by the nonlinear differential-mode gain. Practical design techniques are illustrated with an example of a real buffer amplifier for ISDN purposes.

The study of DACs and ADCs is limited to the only relevant technique for DSP applications with a high dynamic range: the oversampled data converters (also denoted as Sigma-Delta data converters). The basic principle of this converter type is explained in **Chapter 3** and compared with other data converter types. It is demonstrated that the oversampling technique allows the realisation of data converters with a high dynamic range and a moderate sampling rate.

In **Chapter 4**, the design of Sigma-Delta A-to-D converters with more than two integrators is discussed. With behavioural simulations, the stability of a Fourth-order ADC is studied. It is shown that a stable modulator can be obtained with a well-considered scaling of the internal signals in the modulator loop. This approach is compared with some alternatives. Design techniques to eliminate the signal degradation due to clock feedthrough and to suppress the spurious coupling between the analog and the digital circuits on one chip are discussed and verified with a realised circuit.

The practical design requirements for oversampled D-to-A converters are studied in **Chapter 5** and compared with classical multi-bit D-to-A converters. With a general calculation technique, the harmonic distortion of several designs is compared. A practical realisation of a current-steering DAC is described in detail.

This book originated from the PhD dissertation of the first author, which describes research carried out in the ESAT-MICAS group of the *Catholic University Leuven*, Belgium. Some circuits realised during this research work are presented here as design examples. Later, several sections were added, describing the state of the art and the publications of other authors. In this way, this book can serve both as a general introduction and as a reference work in the fields of low-distortion analog circuits and oversampled data converters. It can also be used for an advanced graduate course covering these topics.

Finally, we wish to thank all persons who have contributed towards the realisation of this book. In particular, P. Meulemans, B. Maes, P. Heyrman, P. Ampe, L. Verdeyen, H. Vandooren, P. Vandeloo, P. Wambacq and G.M. Yin, who have contributed on the research results described in this book. We are also grateful to the Belgian IWONL, to Alcatel Bell and to Mietec Alcatel for their support and for the many useful technical discussions.

Frank Op 't Eynde
ASIC Design Center
Mietec Alcatel
Brussels, Belgium

Willy Sansen
Department of Electrical Engineering
Katholieke Universiteit Leuven
Leuven, Belgium

1   THE POWER CONSUMPTION OF CMOS WIDEBAND AMPLIFIERS

1.1. INTRODUCTION: WHY <u>CMOS</u> HF AMPLIFIERS?

An operational amplifier is a key building block in various analog functions. In a broad class of analog circuits with a high dynamic range, the signal degradation is directly related to the limited gainbandwidth (GBW) of the amplifiers. For example, increasing the GBW of an amplifier increases the loop gain and therefore the distortion performance. Therefore, low-distortion amplifiers require a GBW which is much larger than the signal frequency. An other example is an oversampled data converter: increasing the GBW allows to increase the Oversampling Ratio which benefits the signal-to-noise ratio. In many applications, amplifiers with a large GBW are required, although the signal frequencies can be relatively small.

Because of the larger transistor cut-off frequencies, a modern bipolar technology is more suited for the implementation of wideband amplifiers than a CMOS technology. However, when the amplifiers have to be integrated with a large digital circuit on one chip, the CMOS technology is advantageous because of its higher integration density.

The main objective of this introductory chapter is to explore the possibilities and limitations of the CMOS technology for the implementation of wideband amplifiers.

Since wideband amplifiers require large transconductances to drive the load capacitances up to high frequencies, they consume a large amount of power. In a real design situation, it is important to budget the power consumption in advance in order to estimate the system cost. For low-frequency amplifiers where the GBW is much smaller than the transistor cut-off frequencies, the power consumtion varies linearly with the amplifier GBW [1]. In this chapter, it is shown that due to the finite transistor cut-off frequencies, the power consumption of high-frequency amplifiers increases with the square of the GBW.

It is well-known [2-4] that the HF characteristics of a MOS device differ from the low-frequency ones. In Section 1.2, the most important MOS characteristics for HF applications are resumed.

The power consumption of HF Operational Transconductance Amplifiers (OTA) is studied in Section 1.3. Scaling laws are derived for the power consumption as a function of the GBW, the load capacitance and the process characteristics. Three basic amplifier structures are compared. It is shown that minor modifications on existing circuits [5-7] can save over a factor of two of power consumption.

In Section 1.4, two design examples of CMOS wideband OTAs are presented, with a GBW of 150 MHz and 800 MHz respectively. The layout, the packaging and the measurement technique are discussed.

1.2. THE HF CHARACTERISTICS OF A MOSFET

There are four important differences between a bipolar transistor and a MOS device that make the latter device less suitable for high-frequency applications:
- For the same current, the transconductance of a MOS device is much smaller than that of a bipolar transistor.
- The cut-off frequency (f_T) of a MOSFET is much smaller than that of a bipolar transistor.
- For the same transconductance, the parasitic capacitances of a MOSFET are larger.
- A MOSFET acts as a delay line. This results in an additional phase shift which degrades the amplifier stability.

In this section, all these effects are discussed and modelled.

1.2.a. The cut-off frequency

Compared with a MOS transistor, the small-signal characteristics of a bipolar device are relatively simple: the transconductance (g_m) is proportional to the current and independent of the device size while the cut-off frequency is constant at first order [8]. A MOS transistor is more complex since the transconductance is a function of both the aspect ratio (W/L) and the drain current (I). In Fig. 1.1, curves of constant g_m are depicted in the I-W/L plane. For small current densities (upper left of Fig. 1.1), the transconductance is independent of the aspect ratio. This is the region of **Weak Inversion** [9]. For large current densities (lower right of Fig. 1.1), g_m is function of the aspect ratio only. In this region of operation, the drain current varies linearly with the gate voltage. This is the **velocity saturation** region [10], or breefly the **Linear** region $^{(*)}$. Between these two regions is the **Strong Inversion** region. Here, the

---

(*) Note: this **Linear** region is not the **triode** region.

transconductance is a function of both the current and the aspect ratio. For simplicity, only the linear approximation (see the bold line in Fig. 1.1.) is considered in this text.

The cut-off frequency of a MOS transistor varies with the operating point. This variation is different for the three regions described above. Therefore, it is worthwhile to investigate the HF MOS characteristics more in detail for each of the three operating regions.

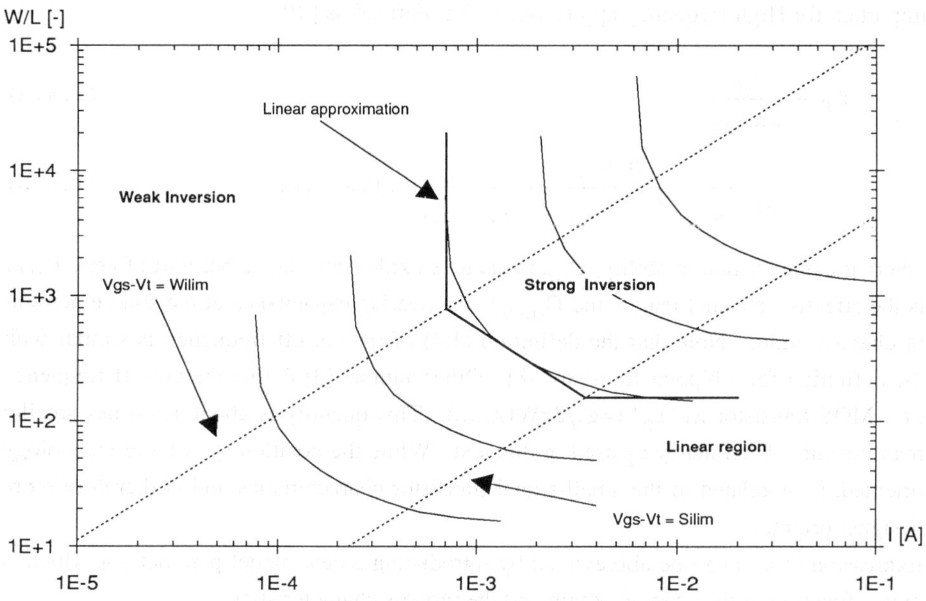

Fig. 1.1: measured transconductance values for a 3 μm nMOS transistor
——: Curves of constant g_m
——: A linear approximation
......: Curves of constant V_{GS} and f_T

i) The transistor model in Strong Inversion (middle of Fig. 1.1.)

For intermediate current densities, the transistor acts in first order as a square-law device:

$$i = \frac{\mu \cdot C_{OX}}{2} \cdot \frac{W}{L} \cdot (v_{GS} - V_T)^2 \tag{1.1}$$

and the transconductance is given by:

$$g_m = \sqrt{2.\mu.Cox.W/L.I} \qquad (1.2)$$

$$= 2.I/(V_{GS}-V_T) \qquad (1.3)$$

As indicated by expression (1.2), the transconductance is a function of both the current and the aspect ratio. The curves of constant transconductance in Fig. 1.1. have a slope of minus one.

The **cut-off frequency** f_T is a first figure of merit for a transistor which is important for High-frequency applications. It is defined as [8]:

$$f_T = \frac{g_m}{2\pi C_{gs}} \qquad (1.4.a)$$

$$= \frac{\mu.C_{ox}}{2\pi.L_{eff}.[2/3.C_{ox}.L_{eff}+C_{gso}]}.(V_{GS}-V_T) \qquad (1.4.b)$$

where μ is the channel mobility, C_{ox} is the gate oxide capacitance per unit of area, L_{eff} is the effective channel length and C_{gso} is the overlap capacitance at the source per unit of channel width. Note that the definition (1.4) of the cut-off frequency is similar with the definition for a bipolar transistor [8]. Other authors [3] define the cut-off frequency of a MOS transistor as "f_o" ($=g_m/2\pi WLC_{ox}$). This quantity is about 1.5 times smaller than the cut-off frequency f_T used in this text. While the notation f_o is more technology oriented, f_T is related to the small-signal transistor characteristics and is therefore more designer oriented.

Expression (1.4.b) can be abbreviated by introducing a new model parameter φ which is only a function of the channel length and the process characteristics:

$$f_T = \varphi.(V_{GS}-V_T) \qquad (1.5)$$

Numerical values are listed in Table 1.1. for transistors with a gate length of 3 μm. Since φ is proportional with L_{eff}^{-2} (see expression (1.4.b)), short channel devices are favourable for HF applications. Note that the value for a nMOS transistor is about two times larger than for a pMOS transistor.

A second figure of merit for a transistor is the ratio **g_m/I**. This quantity expresses how efficient the drain current is used to generate a transconductance. In Strong Inversion, this ratio is given by:

$$g_m/I = 2/(V_{GS}-V_T) \qquad (1.6)$$

	nMOS	pMOS
$L_{nominal}$	3.0 μm	3.0 μm
L_{eff}	2.56 μm	2.3 μm
μ	600 cm^2/Vs	220 cm^2/Vs
C_{ox}	0.8 fF/μm^2	0.8 fF/μm^2
C_{gso}	0.18 fF/μm	0.28 fF/μm
φ	1.3 GHz/V	600 MHz/V
Wilim	80 mV	80 mV
Silim	0.9 V	1.3 V
α_{gd}	0.1	0.15
α_{db}	0.18	0.8
α_{sb}	0.18	0.8

Table 1.1: 3μm MOSFET Characteristics
The last six parameters are extracted from measurements

As can be seen from expressions (1.5) and (1.6), the cut-off frequency increases linearly with (V_{GS}-V_T) while g_m/I varies inversely proportional with (V_{GS}-V_T). There is an optimum (V_{GS}-V_T) for every transistor where the trade-off between f_T and g_m/I yields optimum characteristics.

ii) The transistor model in Weak Inversion (Upper left of Fig. 1.1)

For low current densities, the drain current varies exponentially with V_{GS} and the transconductance is independent of the aspect ratio [9]:

$$I = Cte.exp(\frac{2.v_{GS}}{Wilim}) \qquad (1.7)$$

and $\quad g_m = 2.I/Wilim \qquad$ when $v_{GS}-v_T < Wilim \qquad (1.8)$

In these expressions, "Wilim" is the **Weak Inversion Limit** which is approximately 80 mV [9]. This can be verified in Fig. 1.1: for low current densities (large W/L and small I), the slope of the curves is vertical.
The ratio g_m/I is constant in Weak Inversion:

$$g_m/I = 2/Wilim \qquad (1.9)$$

The cut-off frequency at the boundary between Weak and Strong Inversion is given by:

$$f_T = \varphi \cdot \text{Wilim} \qquad\qquad (1.10)$$

When V_{GS}-V_T decreases below Wilim, the drain current and the transconductance decrease exponentially according to expressions (1.7) and (1.8). The cut-off frequency decreases according to (1.4.a.) while the ratio g_m/I remains constant. Hence, the trade-off between g_m/I and f_T will never yield a V_{GS} below Wilim. In practice, Wilim is a minimum bound for (V_{GS}-V_T). Its value is shown in Table 1.1.

In Weak Inversion, the ratio g_m/I is maximum but the cut-off frequency is small.

iii) The transistor model in the "Linear Region" (lower right of Fig. 1.1.)

Due to mobility degradation by vertical fields or due to velocity saturation [10], the transistor drain current varies approximately linearly with the gate voltage at high current densities. A source resistance also tends to linearise the transistor I-V relation at higher current densities. As a result, the transconductance is independent of the drain current.

This can be verified in Fig. 1.1: for V_{GS} larger than the "Strong Inversion Limit **Silim**" (large I and small W/L), the curves of constant transconductance are horizontal. This implies that the transconductance is independent of the current and a function of the aspect ratio only. In this region of operation, the cut-off frequency is maximum but g_m/I is very small.

It is not advantageous to bias a transistor with V_{GS}-V_T larger than Silim: the current increases without an increase of g_m or f_T. This implies in practice that Silim is an upper bound for V_{GS}-V_T. Its value is also shown in Table 1.1.

iv) The global transistor model

Fig. 1.2. shows the cut-off frequency of a 3 μm nMOS transistor as a function of V_{GS}, obtained from S-parameter measurements [11-12]. The cut-off frequency of a 3 μm MOSFET is much smaller than that of a modern HF bipolar transistor which is in the order of 10 to 20 GHz [13]. As explained before, the cut-off frequency increases about linearly with V_{GS}, according to expression (1.2), but becomes almost constant at high values.

Measured values for g_m/I as a function of V_{GS} are also depicted in Fig. 1.2. for a 3μm nMOS transistor. As stated earlier, this ratio has a maximum value in the Weak Inversion region. In Strong inversion, it is inversely proportional with (V_{GS}-V_T), according to expression (1.6). A comparison of this curve with the value of 40 V^{-1} for a bipolar transistor clearly shows that a MOSFET requires more power to obtain a certain transconductance value. This is especially true when the cut-off frequency needs to be high (see Fig. 1.2.).

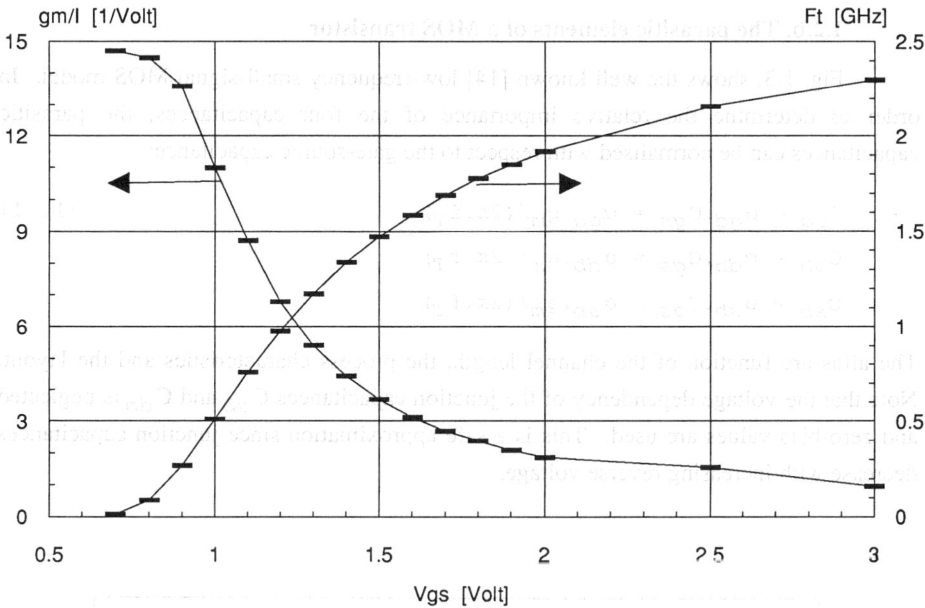

Fig. 1.2: The measured cut-off frequency and g_m/I for a
$3\mu m$ NMOS transistor

Since g_m/I reaches a maximum value in the Weak Inversion region, low-frequency amplifiers require a minimal power consumption when the input transistors operate in Weak Inversion. For high-frequency amplifiers however, the parasitic capacitances are important and transistors with a sufficient cut-off frequency are required. The Weak Inversion region is not necessarily the optimum choice.

In reference [1], a Current Excess Factor is defined which compares the power consumption of an amplifier with that of a CMOS invertor with transistors in Weak Inversion. It is a useful quality factor for low-frequency amplifiers. For high-frequency amplifiers on the other hand, the Weak Inversion region is not the optimum operating region since the cut-off frequency is too small and the parasitic capacitances are too large.

1.2.b. The parasitic elements of a MOS transistor

Fig. 1.3. shows the well-known [14] low-frequency small-signal MOS model. In order to determine the relative importance of the four capacitances, the parasitic capacitances can be normalised with respect to the gate-source capacitance:

$$C_{gd} = \alpha_{gd} \cdot C_{gs} = \alpha_{gd} \cdot g_m / (2\pi \cdot f_T)$$

$$C_{db} = \alpha_{db} \cdot C_{gs} = \alpha_{db} \cdot g_m / (2\pi \cdot f_T)$$

$$C_{sb} = \alpha_{sb} \cdot C_{gs} = \alpha_{sb} \cdot g_m / (2\pi \cdot f_T)$$

$$(1.11)$$

The alfas are function of the channel length, the process characteristics and the layout. Note that the voltage dependency of the junction capacitances C_{sb} and C_{db} is neglected and zero-bias values are used. This is a safe approximation since junction capacitances decrease with increasing reverse voltage.

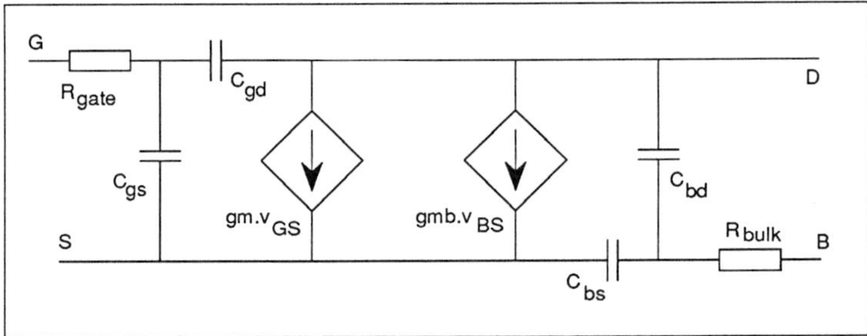

Fig. 1.3: The low-frequency small-signal equivalent of a
MOSFET

For high-frequency applications, the parasitic capacitances should be as small as possible. In order to obtain this goal, special precautions must be taken in the layout. Fig. 1.4. shows three different layout styles. The corresponding values for α are collected in Table 1.2. Compared with the **straightforward layout** of Fig. 1.4.a, the source and drain areas of the **interdigitated layout** are 40% smaller. This results in smaller junction capacitances. With the **waffle layout** of Fig. 1.2.c, the junction capacitances are even smaller. However, the polysilicon squares at the gate crossings (see Fig. 1.4.c.) do not contribute to the transistor width although they contribute to the gate capacitance. Moreover, the gate resistance is larger for this waffle layout style.

	straight-forward layout	interdi-gitated layout	waffle layout
nMOS			
α_{gd}	0.1	0.1	0.1
α_{db}	0.517	0.18	0.1
α_{sb}	0.517	0.18	0.1
pMOS			
α_{gd}	0.15	0.15	0.15
α_{db}	1.57	0.8	0.44
α_{sb}	1.57	0.8	0.44

Table 1.2: The parasitic capacitances of 3 μm MOS transistors for different layout styles

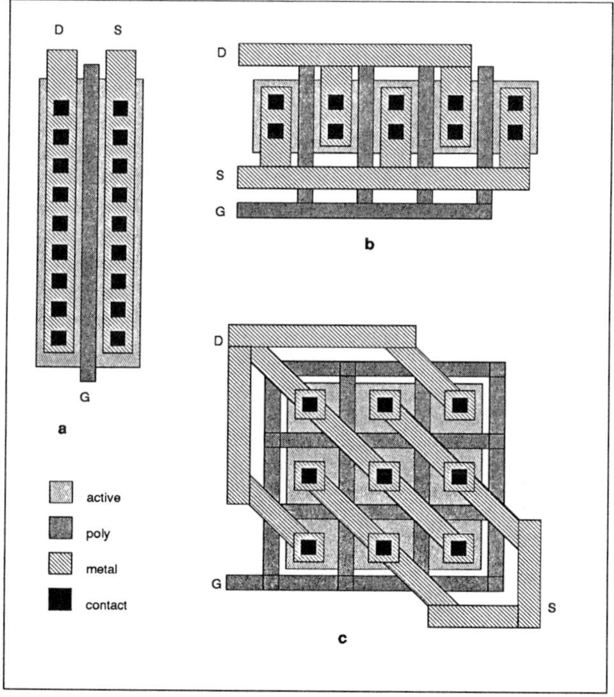

Fig. 1.4: a) a straightforward layout,
b) an interdigitated layout,
c) a waffle layout

Compared with a modern bipolar transistor, the parasitic capacitances of a 3 μm nMOS device are considerably larger. For example, for a 3 μm nMOS transistor with a transconductance of 10^{-2} S and a cut-off frequency of 1 GHz, C_{gs}, C_{gd}, C_{db} and G_{sb} respectively equal 1.6 pF, 0.17 pF, 0.5 pF and 0.5 pF. For a state-of-the-art NPN bipolar transistor on the other hand, the cut-off frequency is typically in the order of 15 GHz [13]. For such a transistor with the same transconductance, C_π, C_μ and C_{cs} have an order of magnitude of respectively 0.1 pF, 30 fF and 30 fF. Note that there is no capacitance between the emitter and the substrate which would be the equivalent of C_{sb}. Clearly, the parasitic capacitances of the bipolar transistor are considerably smaller than those of the 3 μm MOSFET.

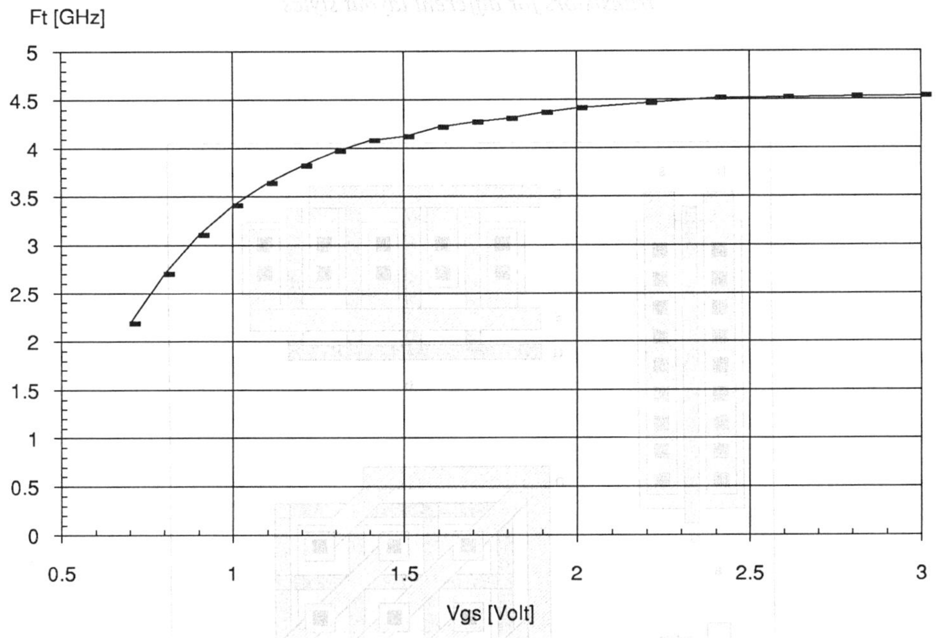

Fig. 1.5: The cut-off frequency of an nMOS with 1.2 μm channel length

This conclusion remains valid when comparing the modern bipolar NPN transistor with a state-of-the-art short-channel nMOS device. In Fig. 1.5, the cut-off frequency of a 1.2 μm nMOS is plotted versus V_{GS}. A comparison of this curve with Fig. 1.2. shows that the cut-off frequency of the 1.2μm nMOS increases more rapidly with increasing V_{GS} than for the 3 μm nMOS. This is in agreement with expression (1.4.b.) and (1.5): φ is proportional with L^{-2}. However, compared with Fig. 1.2, the curve of Fig. 1.5. reaches a maximum at a smaller V_{gs}. The maximum cut-off frequency for the 1.2 μm nMOS is only about a factor of two larger than for the 3 μm device. This cut-off frequency is still considerably smaller than the f_T of the bipolar transistor. As a result, the parasitic capacitances of the 1.2 μm nMOS are larger than those of a state-of-the-art bipolar transistor with the same transconductance.

1.2.c. The delay-line effects in a MOSFET

Fig. 1.6 shows a cross-section of a MOSFET. When the voltage at the gate contact changes, it takes some time to modify the channel charge distribution and the drain current: the transconductance of a MOSFET shows an intrinsic delay. This delay consists of two components:

1) The gate resistor forms a lossy distributed delay line with the gate capacitance. When the voltage at the gate contact changes, the gate signal needs some time to travel along the gate width. The delay time can be calculated as [12]:

$$\tau_g = 1/3 * R_{gate} \cdot C_{gate}$$
$$\approx 0.22 * W^2 \cdot \rho \cdot C_{ox} \tag{1.12}$$

where W is the transistor width, ρ is the gate sheet resistance and C_{ox} is the gate oxide capacitance per unit of area. For example for a MOSFET with a width of 100 μm, C_{ox} being 1 fF/μm^2 and ρ being 20 Ohm, τ_g is 42 psec. For a signal frequency of 300 MHz, this causes an additional phase shift $(2\pi f.\tau)$ of 4.5°. This effect degrades the phase margin of HF amplifiers. Since the delay is proportional with W^2, it can be concluded that a MOS transistor for HF applications not only needs to be short but it should be narrow as well. Wide transistors have to be split up in narrows sections. Note that a base resistance causes a similar effect in bipolar transistors [8].

2) After the gate signal has arrived at the transistor, it takes time to modify the channel charge distribution. More elaborated HF MOSFET models describe this effect with non-reciprocal capacitances [2-3]. In [3], it is shown that this is mathematically equivalent to a description with a frequency dependent transconductance and bulk-transconductance. An expression for the frequency dependent g_m can be calculated from a nonquasistatic transistor model [3]:

$$g_m(j\omega) = \frac{g_m(DC)}{1 + 0.4 . j\omega/2\pi f_T + 0.05 . (j\omega/2\pi f_T)^2 + \dots} \tag{1.13}$$

For frequencies up to the cut-off frequency, the magnitude of this expression varies with only three percent, so it can be considered as being constant. In this frequency range, the phase varies linearly and the delay time equals:

$$\tau_c = 0.4/2\pi f_T = 0.064/f_T \qquad (1.14)$$

For example for a cut-off frequency of of 2 GHz, τ_c equals 32 psec. For a signal frequency of 300 MHz, the additional phase shift is 3.5°.

Note that none of these delay effects are modelled in the SPICE MOS models.

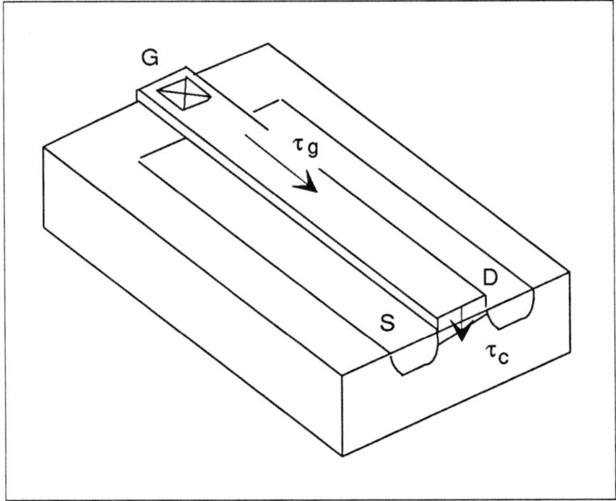

Fig. 1.6: The delay mechanism in a MOS transistor

1.2.d. Conclusions

The delay effects are the major difference between the low-frequency SPICE models and the high-frequency transistor behaviour. The accuracy of the SPICE simulations can be improved by explicitly including a gate resistor for every transistor. From expressions (1.12) and (1.14), it can be seen that this resistor has to equal

$$R_{simulation} = 1/3*R_{gate} + 0.4/g_m \qquad (1.15)$$

The phase shift due to the pole formed by this resistor and C_{gs} approximates the effects due to the distributed gate resistance (see expression (1.12)) and the channel delay (see equation (1.14)).

HF amplifiers require MOS devices with a large transconductance and a large cut-off frequency. From Fig. 1.2, it can be concluded that the power consumption will be large.

1.3. POWER MINIMISATION OF WIDEBAND OTAS

In this section, scaling laws are derived for the power consumption of CMOS HF single-stage amplifiers.

In Fig. 1.7.a, the most simple CMOS OTA with an active load is depicted [14]. This circuit is not suited for high-frequency applications because the gate-drain capacitance of M_1 is connected between the input node and the output node. Due to the Miller-effect, this capacitance forms a low-frequency pole with the gate resistance of M_1, resulting in an additional phase shift that degrades the amplifier stability. This problem is eliminated in the "Current Gain amplifier" [5] of Fig. 1.7.b. First, this amplifier type is optimised towards a minimum supply current consumption for a given GBW and load capacitance. This current is compared with the current consumption of a "Folded-Cascode" amplifier. It is demonstrated that the Folded-Cascode schematic is more suited for amplifiers with a large gainbandwidth. With a modification of the Folded-Cascode schematic, the HF characteristics can be further improved yielding a considerably smaller power consumption.

1.3.a. Power minimisation of a Current Gain OTA

In Fig. 1.7.b, a Current Gain amplifier [5] is depicted. The GBW and the total supply current are given by:

$$GBW = \frac{g_{m1} \cdot B}{2\pi C_L} = \frac{2 \cdot I_{M1}}{(V_{GS} - V_T)_1} \cdot \frac{B}{2\pi C_L} \tag{1.16}$$

$$I_{supply} = 2 \cdot (1+B) \cdot I_{M1} \tag{1.17}$$

where B is the current mirror gain factor and I_{M1} is the current through transistor M_1.

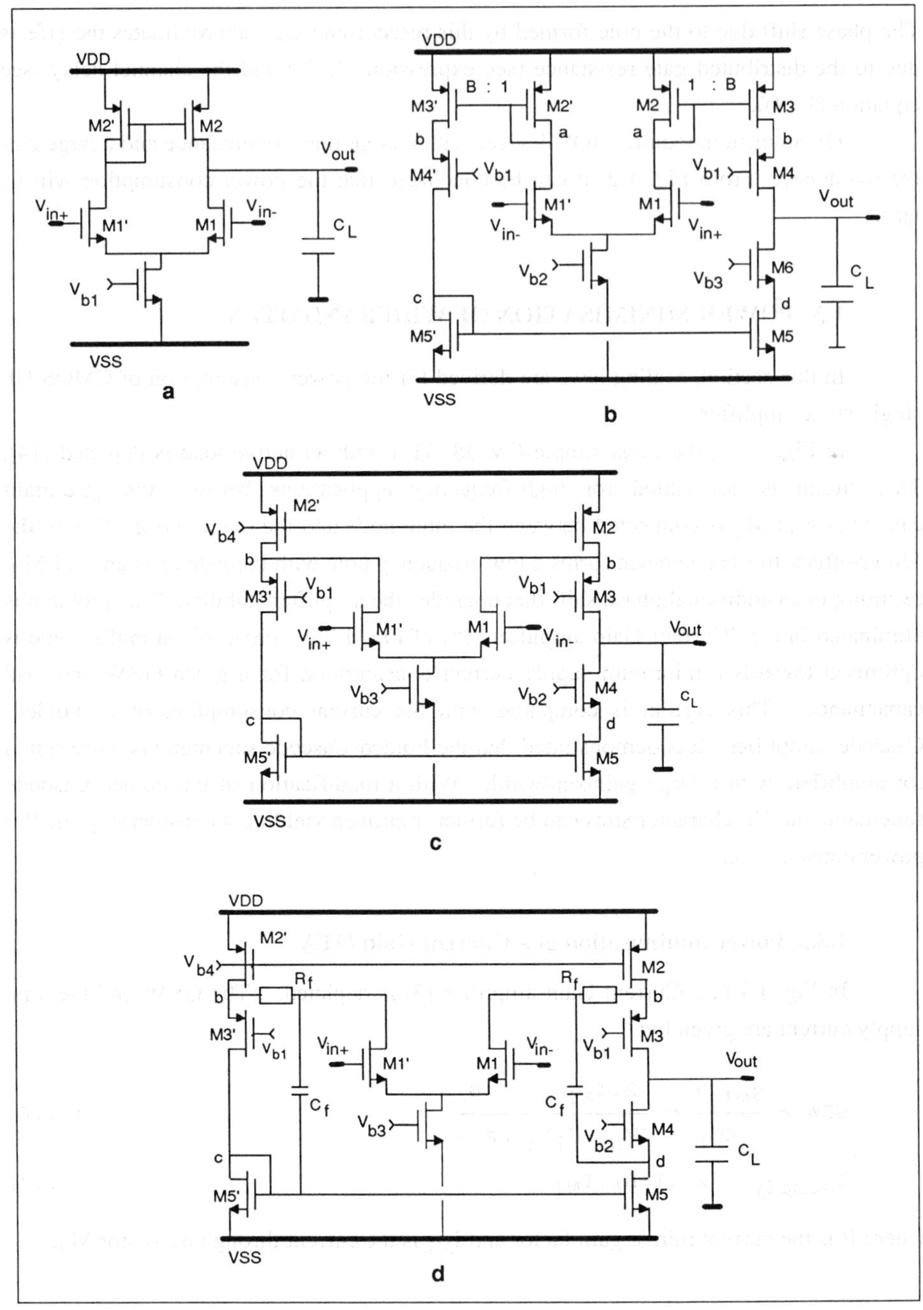

Fig. 1.7: a) a simple CMOS OTA
b) a Current Gain amplifier
c) a Folded-Cascode amplifier
d) a Modified Folded-Cascode amplifier

Expression (1.16) indicates that the GBW can be increased indefinitely by increasing g_{m1}, thus by increasing the bias current and the aspect ratios of the input transistors. However, besides a sufficient GBW, most applications require a minimum value for the phase margin. The phase margin determines the (undesired) resonance peaking in the amplifier frequency response characteristic and the damping of the transient response [15-16]. A phase margin of at least 45° is necessary for most applications. Hence, sufficiently high values for the internal poles at nodes a to d in Fig. 1.7.b. are required.

With a constant V_{GS} for transistors M_3 and M_4, the pole on node b in Fig. 1.7.b. is independent of the current through M_3 and M_4: increasing this current requires larger transistors, thus g_{m4} and the parasitic capacitances increase together. The only way to increase this pole is to increase V_{GS} of M_3 and M_4 as far as possible.

For the poles on nodes c and d, the situation is similar: these poles are related to the transistor cut-off frequencies and to the alfas in (1.11) but not to the currents.

The pole on node a is the first non-dominant pole. It is given by:

$$p_a = \frac{g_{m2}}{2\pi \ \{ \ C_{gd1} + C_{db1} + C_{gs2} \cdot (1+B) + C_{db2} + M \cdot C_{gd3} \ \}} \qquad (1.18)$$

The subscripts 1 , 2 and 3 stand for the transistor numbers in Fig. 1.7.b. M is the Miller factor of C_{gs3}. It approximately equals two when the aspect ratio of M_3 equals that of M_4. This pole can be increased by increasing the V_{GS} of transistors M_1 and M_2 or by decreasing the current mirror gain B.

A sufficient phase margin can be obtained only with a sufficient value for the second pole p_a.

As can be seen from expressions (1.16), (1.17) and (1.18), there are four design variables $((V_{GS}-V_T)_1$, $(V_{GS}-V_T)_2$, B and $I_{M1})$ which influence the GBW, the second pole and the supply current. In a real design situation, the determination of these variables is a complex trade-off between conflicting requirements such as power consumption, GBW, noise requirements, input capacitance, input common-mode range or output swing. In this chapter, the problem is limited to a power minimisation for a given GBW, load capacitance and second pole.

Three operating conditions can be distinguished:

i) M_1 and M_2 operate in Strong Inversion

With the notations introduced in section 1.2, expression (1.18) can be rewritten as:

$$P_a = \frac{2 \cdot I_{M1} / (V_{GS} - V_T)_2}{\dfrac{2 \cdot I_{M1} \cdot (\alpha_{db1} + \alpha_{gd1})}{(V_{GS} - V_T)_1^2 \cdot \varphi_1} + \dfrac{2 \cdot I_{M1} \cdot [1 + \alpha_{db2} + B(1 + M\alpha_{gd2})]}{(V_{GS} - V_T)_2^2 \cdot \varphi_2}} \qquad (1.19)$$

From formulas (1.16), (1.17) and (1.19), one huge equation can be derived, expressing I_{supply} as a function of the GBW, the second pole, the load capacitance and the four design variables mentioned before. It is easy to calculate that I_{supply} is minimum when $(V_{GS} - V_T)_1$, $(V_{GS} - V_T)_2$ and B are chosen according to:

$$(V_{GS} - V_T)_1 = \frac{1}{\varphi_1} \cdot 2 \cdot P_a \cdot \sqrt{[1 + \alpha_{db2} + B(1 + M\alpha_{gd2})] \cdot [\alpha_{db1} + \alpha_{gd1}] \cdot \varphi_1 / \varphi_2} \qquad (1.20)$$

$$(V_{GS} - V_T)_2 = \frac{1}{\varphi_2} \cdot 2 \cdot P_a \cdot [1 + \alpha_{db2} + B(1 + M\alpha_{gd2})] \qquad (1.21)$$

and $\quad B = \dfrac{1}{2} \left\{ 1 + \sqrt{1 + 8 \cdot \dfrac{1 + \alpha_{db2}}{1 + M \cdot \alpha_{gd2}}} \right\} \approx 2 \qquad (1.22)$

The supply current then equals

$$I_{supply} = 4\pi \cdot GBW \cdot C_L \cdot (1 + 1/B) \cdot P_a \cdot \sqrt{\frac{[1 + \alpha_{db2} + B(1 + M\alpha_{gd2})] \cdot [\alpha_{db1} + \alpha_{gd1}]}{\varphi_1 \cdot \varphi_2}} \qquad (1.23)$$

This expression indicates that the power consumption increases linearly with the GBW, with the load capacitance and with the second pole. However, when increasing the GBW, the second pole has to increase as well to maintain a sufficient phase margin. For a constant ratio between the second pole and the GBW, the power consumption increases with the square of GBW. This can be explained as follows: the internal poles of the open loop gain are functions of the internal parasitic capacitances, thus of the transistor cut-off frequencies. When increasing the GBW, transistors with a higher f_T are required to ensure stability. According to expression (1.5), this implies that $(V_{GS} - V_T)_1$ has to increase. Hence, I_{M1} / g_{m1} increases (see expression (1.3)). Since g_{m1} itself has to increase proportional with the GBW according to equation (1.16), the supply current increases with the square of GBW.

With the numerical values of Table 1.1, it can be calculated from expression (1.23) that a realisation with pMOS input transistors requires about 70% more power compared with the complementary equivalent with nMOS input transistors. Consider for instance a design example where a GBW of 50 MHz is required for a load capacitance of 10 pF and a second pole of 150 MHz. According to expression (1.23), a realisation with nMOS input transistors, as drawn in Fig. 1.7.b, requires a supply current of 1.78 mA while a complementary equivalent would consume 2.95 mA. Hence, when all the transistors operate in Strong Inversion, a realisation with nMOS input transistors is favourable.

ii) M_1 operates in "Weak Inversion"

For small values of the second pole, $(V_{GS}-V_T)_1$ can be small (see expression (1.20)). When $(V_{GS}-V_T)_1$ is below "Wilim", I_{M1}/g_{m1} is a constant (see expression (1.7)). Expressions (1.16) and (1.19) become now:

$$GBW = \frac{g_{m1} \cdot B}{2\pi C_L} = \frac{2 \cdot I_{M1}}{Wilim} \cdot \frac{B}{2\pi C_L} \tag{1.24}$$

$$P_a = \frac{2 \cdot I_{M1}/Wilim}{\dfrac{2 \cdot I_{M1} \cdot (\alpha_{db1}+\alpha_{gd1})}{Wilim^2 \cdot \varphi_1} + \dfrac{2 \cdot I_{M1} \cdot [1+\alpha_{db2}+B(1+M\alpha_{gd2})]}{(V_{GS}-V_T)_2^2 \cdot \varphi_2}} \tag{1.25}$$

In a similar way as for the previous case, the supply current can be calculated from expressions (1.7), (1.24) and (1.25). As long as M_1 is in Weak Inversion, the amplifier supply current varies linearly proportional with the GBW. This is in agreement with the results obtained in reference [1]. However, in this reference, the Weak Inversion region is presented as the optimum operating region for all CMOS amplifiers. In reality, the Weak Inversion region is an optimum choice, only for low-frequency amplifiers. From expression (1.20), it can be seen that for the amplifier of Fig. 1.7.b, the input transistors operate in Weak Inversion when the second pole is below 47 MHz. For the complementary circuit, the optimum operating point for the input transistors is in the Weak Inversion region when a second pole below 19 MHz is required. For applications with a second pole at higher frequencies, an amplifier with the input transistors biased in Strong Inversion will consume less power.

iii) M_1 and M_2 operates in the "Linear Region"

The larger the second pole, the larger $(V_{GS}-V_T)_1$ and $(V_{GS}-V_T)_2$ have to be (see expressions (1.20) and (1.21)). When these reach "Silim", no further improvement of the cut-off frequencies can be obtained. As can be seen from (1.19), when $(V_{GS}-V_T)_1$ and $(V_{GS}-V_T)_2$ have reached their maximum values, the second pole can only be increased a little further by decreasing the current mirror factor. The supply current will increase dramatically.

iv) The power consumption as a function of the GBW

According to expression (1.16), the transconductance of an OTA (g_{m1}.B for the circuit of Fig. 1.7.b.) is proportional with the GBW and with the load capacitance. In Fig. 1.8, the **total amplifier supply current per unit of amplifier transconductance**, obtained from expression (1.23), is depicted versus the second pole (see the curve marked with "Current Gain amplifier - nMOS input"). For a second pole below approximately 150 MHz for the 3 μm design, this curve increases linearly with the second pole.

The curve for the complementary schematic with pMOS input transistors is also depicted. For a second pole below 250 MHz (see the crossing of the two dashed curves in Fig. 1.8), a realisation with nMOS input transistors as depicted in Fig. 1.7.b. is preferable over the complementary circuit. For a second pole at higher frequencies, a complementary realisation with pMOS input amplifiers requires less power.

For a second pole beyond 200 MHz, the power consumption starts to grow very quickly. This illustrates that due to technological limitations, a GBW higher than about 50 MHz is not feasible for this amplifier type in this technology.

Equation (1.23) expresses the supply current for the optimised design. In reality, other considerations such as noise performance, input common mode range or output swing can result in a design which is not optimum in terms of power consumption. Since the resulting supply current will be larger, expression (1.23) estimates the **minimum** required power consumption of a Current Gain OTA.

1.3.b. The power optimisation of a Folded-Cascode OTA

In Fig. 1.7.c, a Folded-Cascode OTA is shown. Compared with the Current Gain amplifier, this structure has no node a and hence, there is one pole less. In the same way as for a Current Gain amplifier, the power consumption of this amplifier type can be analysed as a function of the GBW, the load capacitance and the second pole on node b. The results are depicted in Fig. 1.8. (see the curves marked with "Folded-Cascode"). In the middle, these curves increase about linearly with the GBW. This implies that, again, the power consumption is proportional with the square of GBW. And again, a realisation with nMOS input transistors is preferable for low-frequency applications while a complementary realisation with pMOS input transistors is more suitable for higher frequencies.

This trade-off can be eliminated by modifying the Folded-Cascode schematic as illustrated in Fig. 1.7.d. [17]. For high frequencies, this circuit acts as an all-nMOS design because the HF current flows through the capacitors C_f. Its supply current per unit of amplifier transconductance is also depicted in Fig. 1.8. As can be seen, this circuit with nMOS input transistors can save over 50% of power compared with the other amplifier types.

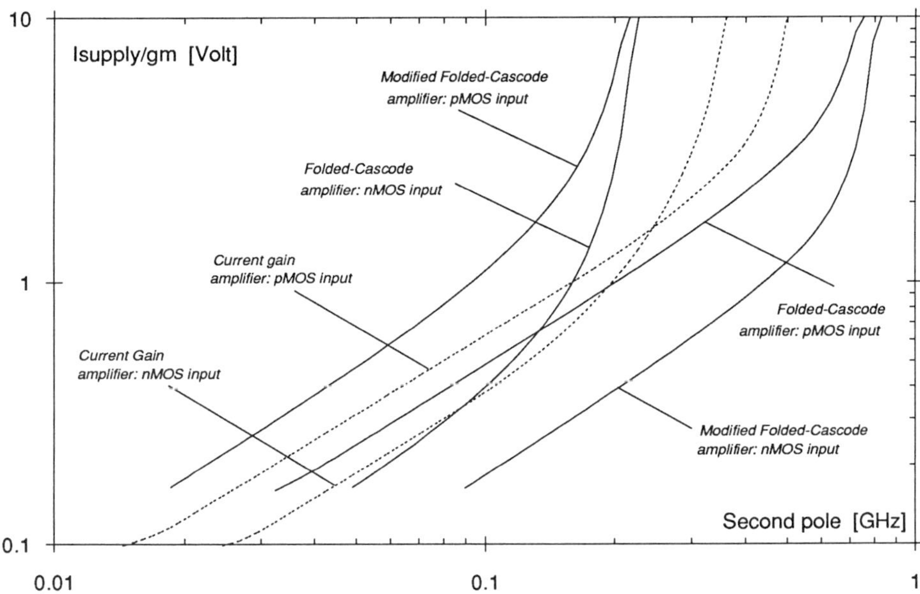

Fig. 1.8: Supply Current per unit of transconductance for different amplifier types in the 3 µm process

1.3.c. An example

For sake of illustration, the estimated power consumption of the three circuits described above, obtained from Fig. 1.8, is compared in Table 1.3. for a GBW of 150 MHz, for a second pole of 600 MHz and for a load capacitance of 3 pF. The transistor characteristics of the 3 µm process are used, both for the circuits as drawn in Fig. 1.7.b. to c. and for their complementary equivalents. For four of the six considered situations, the requirements are impossible to obtain. The supply current for the circuit of Fig. 1.7.d. is over a factor of two smaller than for the other circuits.

	nMOS input transistors	pMOS input transistors
Current Gain	impossible	impossible
Folded-Cascode	impossible	13.6 mA
Modified Folded-Cascode	6 mA	impossible

Table 1.3: An Example: the supply current for a GBW of 150 MHz, a second pole of 600 MHz and a load capacitance of 3 pF

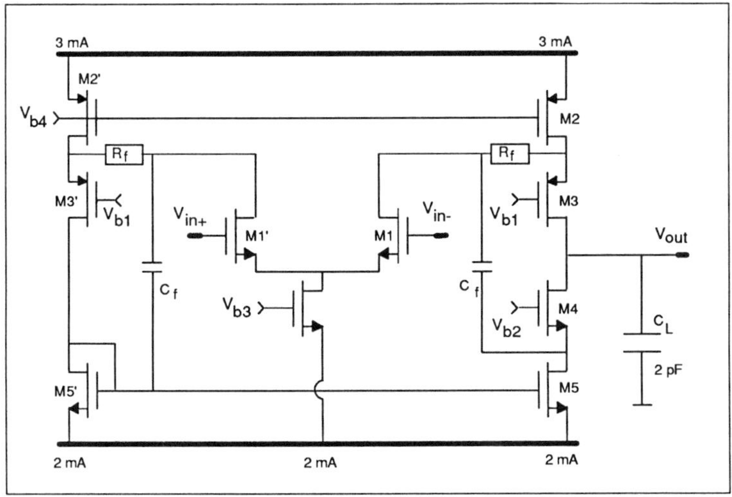

Fig. 1.9: Transistor schematic of the realised 150 MHz amplifier

1.4. PRACTICAL REALISATIONS AND EXPERIMENTAL RESULTS OF THE TWO HF AMPLIFIERS

In order to further illustrate the possibilities and the limitations of the CMOS technology for HF amplifier design, two realised OTAs are presented here. The first is implemented in a 3 μm CMOS technology and achieves a GBW of 150 MHz. The second realisation has a GBW of 800 MHz in a 1.2 μm technology. For both amplifiers, the GBW is about the maximum value that can be obtained in their technology.

1.4.a. A 150 MHz amplifier in a 3 μm CMOS technology

A prototype of the modified Folded-Cascode amplifier of Fig. 1.7.d. is realised in a 3 μm double metal double poly n-well CMOS process [18-19]. It has a GBW of 150 MHz and a phase margin of 58° for a load capacitance of 2 pF. The second pole is designed at 600 MHz which is about the maximum value that can be obtained in this technology (see Fig. 1.8.). The transistor dimensions and currents are collected in Fig. 1.9.

Fig. 1.10. shows a microphotograph of the chip. For the most critical transistors, a waffle layout is applied. The others are realised with an interdigitated layout. In order to obtain a good bulk connection even for high frequencies and to prevent latchup problems, the transistors are surrounded by bulk contacts. In this way, the bulk resistance is minimised. In order to reduce the parasitic source and drain series resistances, multiple diffusion contacts are provided.

The amplifier gain can be derived from scattering parameter measurements [3] [11]. In order to obtain proper 50 Ohm connections between the chip and the S-parameter test set, a dedicated packaging technique has to be developped: the amplifier chip is placed on a thick-film substrate with 50 Ohm striplines for the HF connections. The supply lines and the bias current are decoupled with high-Q SMD capacitors placed as close to the chip as possible. This substrate is mounted on a metal frame as shown in Fig. 1.11. SMA connectors are applied for the HF connections. Dummy chips with open and shorted bonding pads, packaged in the same way serve as calibration standards for the S-parameter test set calibration.

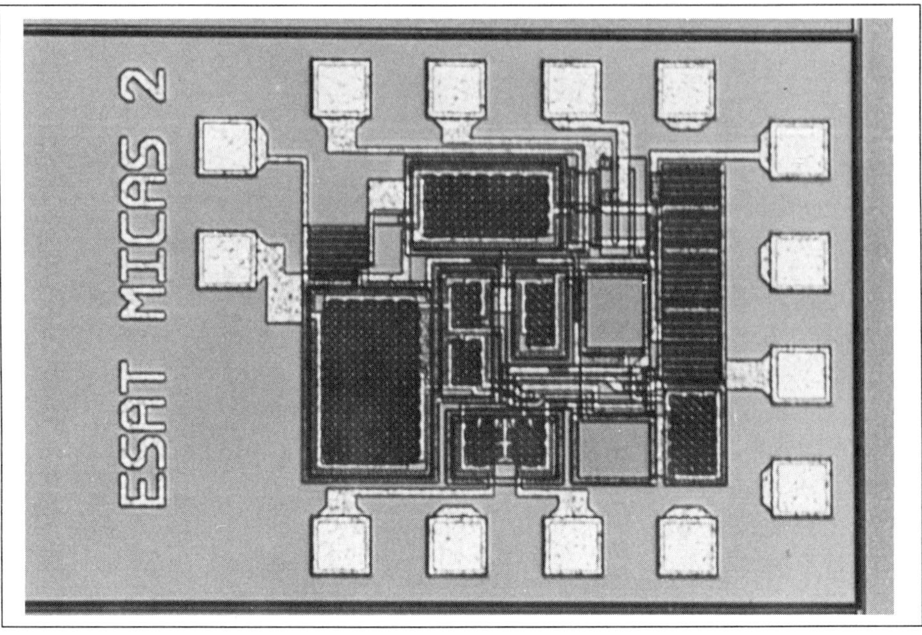

Fig. 1.10: a microphotograph of the 150 MHz amplifier

Fig. 1.11: The HF packaging

After converting the S-parameters to Y-parameters [11], the open loop gain can be derived from

$$A(\omega) \; = \; - \; \frac{Y_{21}(\omega)}{Y_{22}(\omega) + Y_L(\omega)} \tag{1.26}$$

where Y_L is the load conductance. The measurement results are depicted in Fig. 1.12. A GBW of 150 MHz and a phase margin of 58° are obtained.

The amplifier characteristics are summarised in Table 1.4. Note that the amplifier GBW is only a factor of 3.5 smaller than the input transistors cut-off frequency.

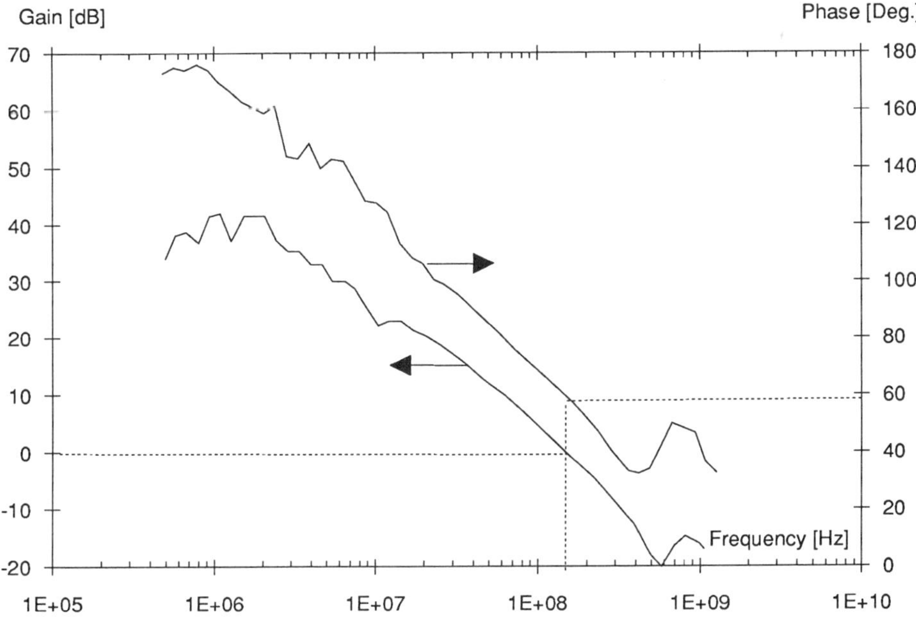

Fig. 1.12: Measurement results for the 150 MHz amplifier

The phase shift at the GBW equals 122°. 90° is due to the dominant pole, 15° is from the second pole at node d in Fig. 1.7.d. and about 7° arises from the pole-zero pair caused by the parasitic capacitance at node c. The remaining 10° is due to the delay line effects in the input transistors as explained in section 1.2.c.

1.4.b. An 800 MHz OTA in a 1.2 µm CMOS technology

The curves of Fig. 1.8. were derived for six different amplifier types in a 3 µm CMOS technology. For CMOS processes with a shorter channel length, these curves shift to higher frequencies.   Fig. 1.13 shows the supply current per unit of transconductance, now for a particular 1.2 µm CMOS technology. The maximum feasible value for the second pole is now about 1.5 GHz.

The operational amplifier of Fig. 1.14, which is a differential version of the "Modified Folded-Cascode" schematic of Fig. 1.6.d, is implemented in this technology [20]. Compared with the single-ended equivalent, the differential schematic eliminates the pole-zero pair due to the nMOS current mirror (M_5-M_5, in Fig. 1.7.d.). Since the phase shift due to this pole-zero pair is avoided, a larger phase shift due to the second pole can be tolerated. A phase margin comparable with that of the previous circuit can be obtained with a second pole of only 2 times the GBW.

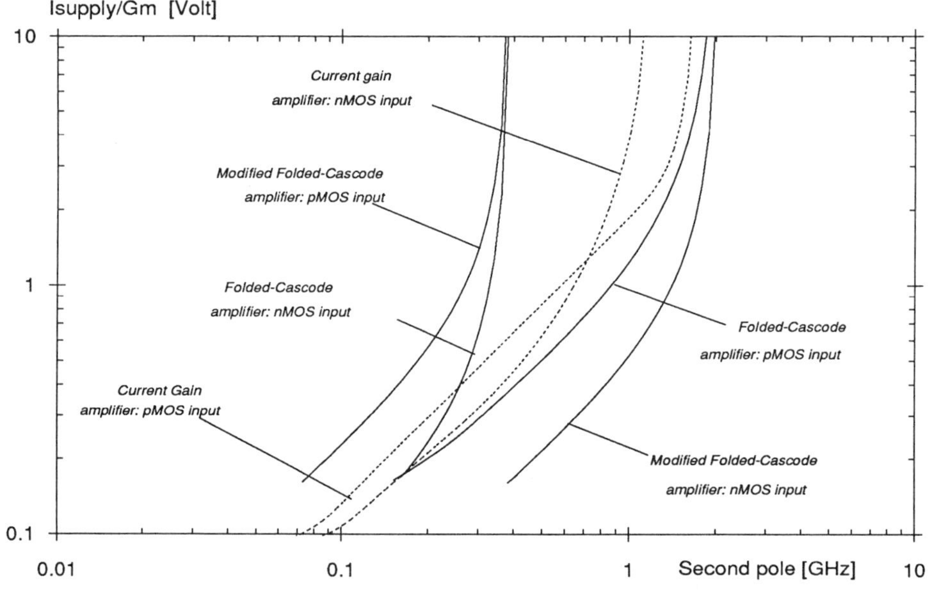

Fig. 1.13: Supply Current per unit of transconductance for different amplifier types in the 1.2 µm CMOS technology

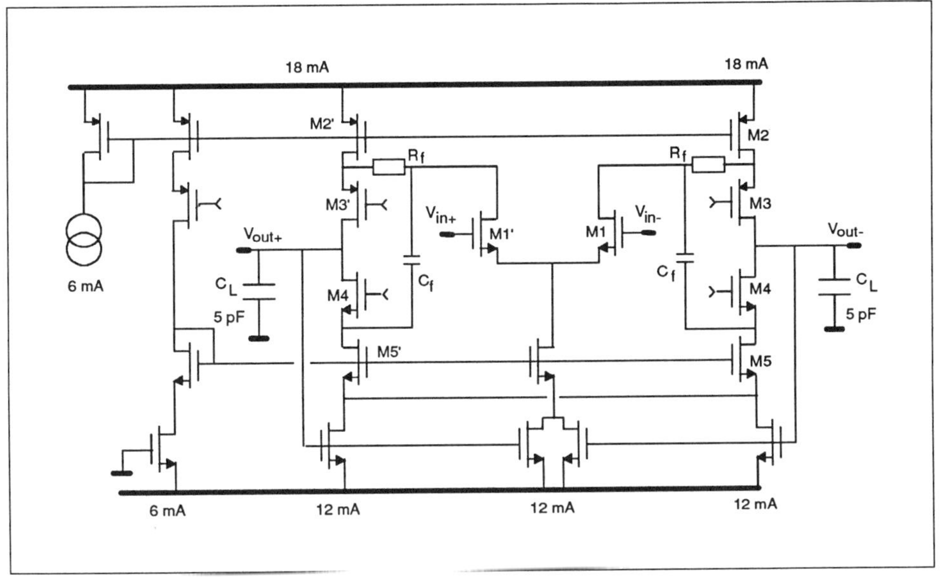

Fig. 1.14: Transistor schematic of the 800 MHz differential amplifier

For high frequencies where the feedforward capacitors can be approximated by short circuits, the ac current flows from M_1 (or M_1') via M_4 (M_4') to the output. Hence, each half of the amplifier schematic reduces to a cascoded single-transistor amplifier. Since this is a minimum configuration with only nMOS transistors, it achieves maximum performance.

Fig. 1.15. shows a microphotograph of the circuit. All the transistors are implemented with an interdigitated layout. In order to minimise the gate resistance, they are splitted up in narrow sections of 50 μm wide. The gate polysilicon fingers are connected to metal at both ends.

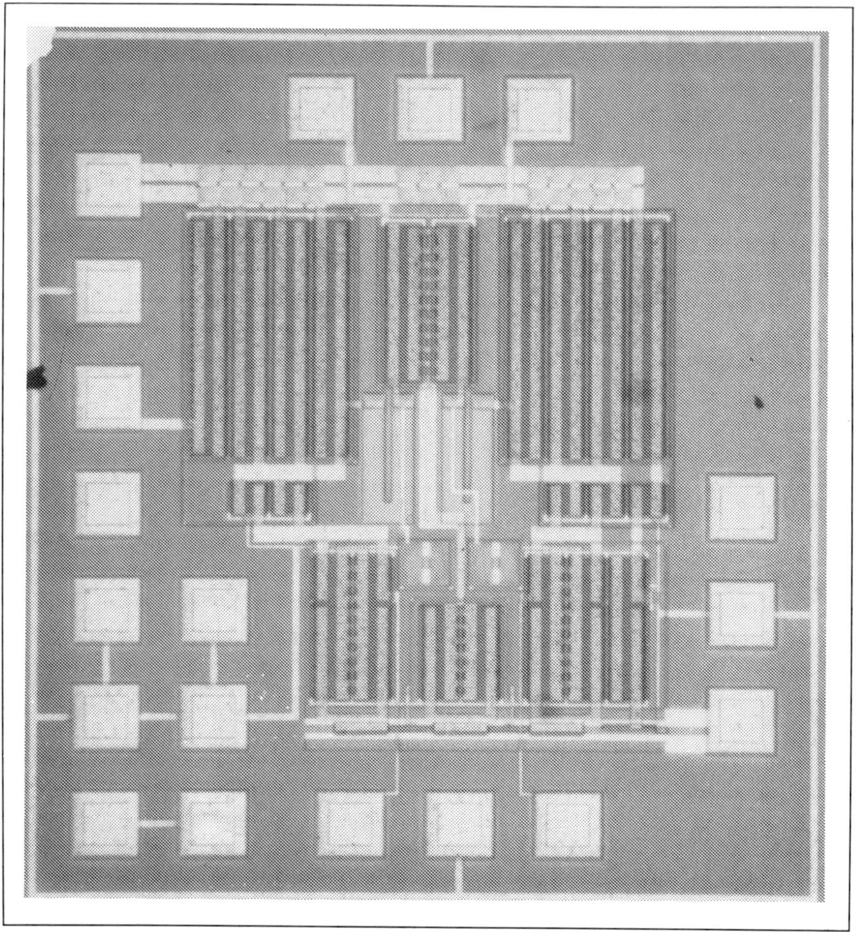

Fig. 1.15: Microphotograph of the 800 MHz amplifier

Fig. 1.16: The HF probes

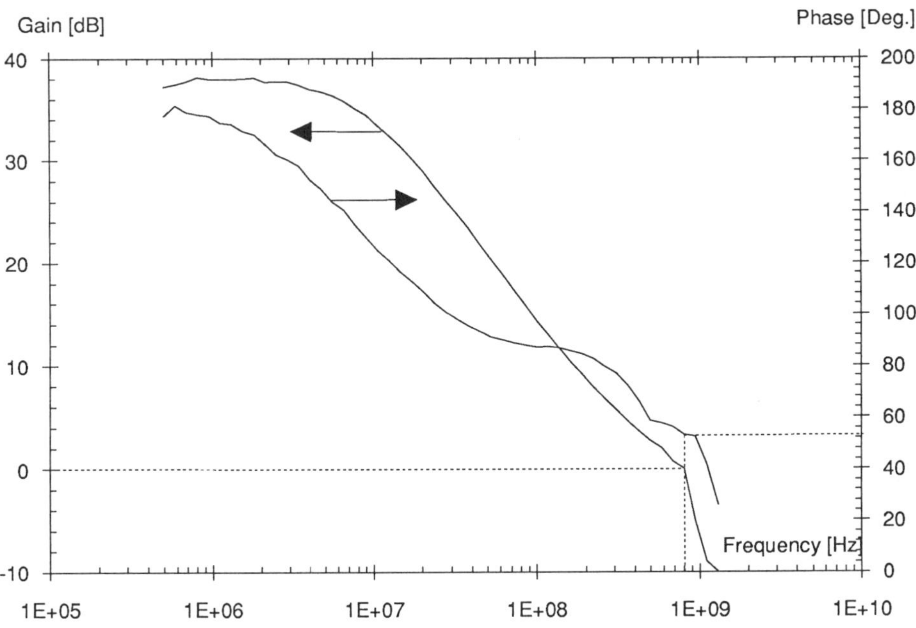

Fig. 1.17: The measured open-loop gain of the 800 MHz differential OTA

In order to minimise the series resistances of the feedforward capacitors, metal contacts are provided all around the capacitor bottom plates. The top plates are on-top contacted with multiple contacts. This amplifier is measured on wafer with dedicated HF probes, shown in Fig. 1.16. They establish a 50 Ohm connection between the S-parameter test set and the amplifier bonding pads. The measurement results are depicted in Fig. 1.17. A GBW of 800 MHz and a phase margin of 53° are achieved for a load capacitance of 5 pF and a power consumption of 240 mW. Compared with the setup of Fig. 1.11, the HF connections of this measurement setup have a higher reproducibility. This results in better measurements, especially for the lowest and the highest frequencies.

The amplifier characteristics are summarised in Table 1.4.

	3 µm amplifier (Fig. 1.9)	1.2 µm amplifier (Fig. 1.14)
V_{DD}, V_{SS}	+/- 2.5 V	+/- 2.5 V
C_L	2 pF	5 pF
Second pole	600 MHz	1.7 GHz
Power Consumption	30 mW	240 mW$^{(*)}$
DC gain	43 dB	39 dB
GBW	150 MHz	800 MHz
phase margin	58°	53°
I_{supply}/g_m	3.1 V	1.4 V
Input Capacitance	0.7 pF	1.7 pF
Output Swing	+/- 1 V	+/- 1 V
Input Offset	3 mV	5 mV
Common-mode output offset	---	< 50 mV

Table 1.4: The amplifier characteristics. The last nine parameters are obtained from measurements

---

(*) 180 mW for the amplifier core and 60 mW for the bias circuit.

1.4.c. Conclusions

The gainbandwidths of 150 MHz and 800 MHz obtained with the examples presented above are far beyond the requirements for most of the applications discussed in next chapters. The huge power consumption on the other hand cannot be tolerated in most practical situations. This illustrates that in most cases, the power consumption is the factor that limits the GBW, rather than the fundamental limitations of the CMOS technology.

Since wideband amplifiers require short devices biased at a large V_{GS}, the DC gain is small. For instance for the discussed examples, the DC gain is only about 40 dB, although cascode transistors are applied. A similar schematic for low-frequency applications would achieve a DC gain of about 80 dB.

1.5. SUMMARY

Compared with low-frequency amplifiers, the design procedure of wideband amplifiers is more complex: the aspect ratio of each transistor is a compromise between a large transconductance and small parasitic capacitances. Hence, the design techniques for low-frequency amplifiers are no longer valid. Where the power consumption of a low-frequency amplifier is proportional with the GBW, the power consumption of a HF amplifier increases with the square of the GBW.

Two design examples of realised wideband amplifiers are presented, one with a GBW of 150 MHz in a 3 μm technology and one with a GBW of 800 MHz in a 1.2 μm technology. Both realisations achieve about the maximum GBW that can be obtained in that particular technology.

It can be concluded that modern CMOS processes allow to realise single-stage amplifiers with a GBW far beyond the requirements for the applications studied in the following chapters of this work. The power consumption however will be large.

1.6. REFERENCES

[1]   M. DEGRAUWE, W. SANSEN: "The Current Efficiency of MOS amplifiers" - *Proc. of ESSCIRC*, 1983, pp. 115-118

[2]   M. BAGHERI, Y. TSIVIDIS: "A Small-Signal dc-to-High-Frequency Nonquasistatic Model for the Four-Terminal MOSFET Valid in All Regions of Operations" - *IEEE Trans. on Electron. Devices* vol. ED-32, No. 11, Nov. 1985, pp. 2383-2391

[3]   P. VANDELOO, W. SANSEN: "Modeling of the MOS Transistor For High-Frequency Analog Design" - Ph. D. thesis K.U.Leuven, 1987

[4]   W. SANSEN, P. VANDELOO: "Modeling the MOS Transistor at High Frequencies" - *Electronic Letters*, vol. 22, No. 15, 17 July 1986, pp. 810-812

[5]   M. MILKOVIC: "Current Gain High-Frequency CMOS Operational Amplifiers": *IEEE J. Solid-State Circuits*, Vol. SC-20 pp. 845-851, August 1985

[6]   R.P. MARTINS, J.E. FRANCA: "A 2.4 µm CMOS Switched-Capacitor Video Decimator with Sampling Rate Reduction from 40.5 MHz to 13.5 MHz" - *Proc. of the IEEE CICC conference*, San-Diego, 1989, pp. 25.4.1-4

[7]   K. BULT, GEELEN: " A Fast-Settling CMOS opamp with 90 dB DC gain and 116 MHz Unity-Gain Frequency" - *Proc. of the IEEE ISSCC*, 1990, pp. 108-109

[8]   I. GETREU: "Modeling the Bipolar Transistor" - Tektronix part. no. 062-2841-00

[9]   E. A. VITTOZ, J. FELLRATH: "CMOS Analog Integrated Circuits based on weak Inversion Operation" - *IEEE J. Solid-State Circuits*, Vol. SC-12 pp. 224-231, June 1977

[10]   S. L. GARVERICK, C. G. SODINI - "Large Scale Linearity of Scaled MOS Transistors" - *IEEE J. Solid-State Circuits*, Vol. SC-22 No. 2, April 1987 pp. 282-286

[11]   G. D. VENDELIN - "Design of Amplifiers and Oscillators by the S-Parameter Method", Wiley, 1982

[12]   F. OP 'T EYNDE - "S-parameter Measurements on MIETEC 3 Micron Transistors": Internal Report K.U.Leuven, 07.02.1989

[13]   H. ISCHINO *et al:* "18 GHz 1/8 Dynamic Frequency divider Using Si Bipolar Technologies" - *IEEE J. Solid-State Circuits* vols. SC-24, no.6, Dec. 1989, pp. 1723-1728

[14]   P. E. ALLEN, D. R. HOLBERG: "CMOS Analog Circuit Design" - Holt, Rinehart and Winston, New York, 1987

[15]   B. Y. KAMATH, R. G. MEYER, P. R. GRAY: "Relationship Between Frequency Response and Settling Time of Operational Amplifiers" - *IEEE J. Solid-State Circuits*, vol. SC-9, No. 6, Dec. 1974 pp. 347-352

[16]   C. CHUANG: "Analysis of the Settling Behaviour of an Operational Amplifier"
         - *IEEE J. Solid-State Circuits*, vol. SC-17, No. 1, Febr. 1982 pp. 74-80

[17]   J. H. HUIJSING, F. TOL: "Monolithic amplifier design with improved HF
         behaviour" - *IEEE J. Solid-State Circuits* vol. SC-11, no. 2, April 1976 pp. 323-
         328

[18]   F. OP 'T EYNDE, W. SANSEN: "Design and Optimisation of CMOS Wideband
         Amplifiers" - *Proc. of the IEEE CICC conference*, San Diego, 1989, pp. 25.7.1-4

[19]   F. OP 'T EYNDE, W. SANSEN: "A 150 MHz OTA in a 3 micron CMOS
         Silicon Technology" *Proc. ISCAS*, Portland, 1989, pp. 86-89

[20]   F. OP' T EYNDE, W. SANSEN: "A 1.2 μm CMOS OTA with 800 MHz GBW"
         - *Proc. of the IEEE CICC conference*, San Diego, 1991

2 LOW-DISTORTION CMOS AMPLIFIER DESIGN

2.1. INTRODUCTION

Besides noise and parasitic coupling of spurious signals, harmonic distortion is the major signal-corrupting factor in high-performance data acquisition systems. As data processing is more and more performed by mixed analog and digital circuits, the distortion specifications for the analog building blocks become more and more severe, in order to fully benefit from the almost unlimited dynamic range of the digital circuits. For example, while the twenty years old Hifi norm required a harmonic distortion of -60 dB, modern audio equipment for compact disks performs at least a dynamic range of a 100 dB. This illustrates the increasing importance of design techniques for low-distortion analog circuits.

The objective of this chapter is to demonstrate that the harmonic distortion of analog circuits is ultimately limited by second-order effects, which are normally neglected in a distortion analysis but become important at low distortion levels. For example, the harmonic distortion of a non-inverting amplifier (e.g. a unity-gain buffer) is normally calculated from the nonlinear behaviour of the open-loop <u>differential-mode</u> gain. However, as will be shown in this chapter, the nonlinear <u>common-mode</u> gain is a second source of distortion which puts strong limits on the achievable distortion performance. In many cases, this second distortion generation mechanism - which is normally neglected - is predominant. As another example, for a power amplifier driving a heavy load, the nonlinear <u>power-supply</u> gain couples power supply variations to the amplifier input. Again, this will cause a harmonic distortion contribution which can be much larger than the distortion due to the nonlinear differential-mode gain. And when the output signal of a power amplifier is fed back to the input stage by thermal coupling, additional distortion is generated which is also neglected in a normal distortion calculation. In this chapter, expressions will be derived to take these second-order effects into account.

Harmonic distortion is caused by circuit nonlinearities: when a nonlinear continuous-time circuit is driven by a sinusoidal signal, parasitic output signals are produced at multiples of the input signal frequency. This signal degradation is referred to as "harmonic distortion". For a discrete-time circuit, circuit nonlinearities can cause intermodulation products between the input signal and the clock signal. Since the clock is a square-wave containing multiple harmonics, this can result in a quasi-continuous output spectrum. (e.g. the quantisation noise in an analog-to-digital converter) This effect is commonly treated as "noise" rather than "distortion". So, not all the effects due to circuit nonlinearities are treated as distortion. Therefore, Section 2.2. briefly defines the term "harmonic distortion", used throughout this work. More details can be found in [1]. Section 2.2. also summarises the standard distortion calculation techniques and the basic distortion expressions used throughout this work.

In Sections 2.3. to 2.5., the influence of the second-order effects mentioned above are investigated. It will be shown that these effects put strong limits on the achievable distortion performance for an integrated amplifier.

The design of a low-distortion amplifier requires various distortion minimisation techniques which are difficult to explain in a general, abstract way. Therefore, these techniques are demonstrated in this book by using a design example: in Section 2.6., a low-distortion Class AB amplifier design example is analysed.

2.2. BASIC DEFINITIONS, TECHNIQUES AND EXPRESSIONS

In order to serve the inexperienced reader, this section recalls the basic theory about harmonic distortion. More details can be found in [1] [2] [3]. Readers who are familiar with distortion calculation techniques can safely omit this section.

2.2.a. Basic distortion definitions

When a sinusoidal signal is applied to a linear, time-invariant system with a single input, the response is a sinusoidal signal with the same frequency and with an amplitude, proportional to the input amplitude.

For nonlinear continuous-time systems and for some nonlinear discrete-time systems on the other hand, the response to a sinusoidal input signal has a frequency spectrum which contains contributions at multiples of the input frequency:

$$\text{if} \quad v_{in}(t) = V_{in} \cdot \cos(\omega_o t) \tag{2.1.a}$$

$$\text{then} \quad v_{out}(t) = \sum_k V_k \cdot \cos(k\omega_o t + \varphi_k) \tag{2.1.b}$$

The output signal contribution at the input frequency is called **the fundamental output signal**. The contribution at the k-th multiple of the input frequency is **the k-th**

harmonic. The **k-th harmonic distortion** is defined as the ratio between the amplitudes of the k-th harmonic and the fundamental signal:

$$HD_k = \frac{V_k}{V_1} \qquad (2.2)$$

The **total harmonic distortion** is defined as:

$$THD = \sqrt{\sum_k HD_k^2} \qquad (2.3)$$

Note that the parameters V_k and φ_k in (2.1.b) depend upon the input frequency and amplitude.

In general, determining the harmonic distortion according to (2.1.b) and (2.2) requires the calculation of Volterra series or even Wiener series [*] [2] [3]. This can be an awkward and time-consuming process. The calculations can however be simplified considerably for circuits satisfying the following two conditions:
1) For the given circuit and the given input signal, the Volterra series [2] reaches convergency.
2) The contributions of the even-order Volterra kernels to the output signal decrease rapidly with increasing order, and so do the contributions due to the odd-order kernels.

A circuit satisfying these conditions is called to operate under **low-distortion conditions**.

2.2.b. Distortion calculation techniques

It is common practice among circuit engineers to calculate harmonic distortion with the Taylor series expansion of the nonlinear DC transfer charcteristic [1]. However, this technique is valid only for **resistive** circuits operating under low-distortion conditions. (A circuit is called resistive when it contains no time-constants: the output reacts immediately upon an input variation). Operational amplifiers are approximately resistive for frequencies well below the dominant pole. Since this pole is typically in the order of 10 Hz, an opamp is usually not a resistive circuit for the frequencies of interest.

For **dynamic circuits** operating under low-distortion conditions, the following calculation technique can be used:

(*) For some more details about Volterra Series and about the Low-distortion conditions, see appendix 2.A at the end of this chapter

The dynamic nonlinear behaviour of a circuit can be characterised mathematically by considering the following - hypothetical - input signal: when the input signal of a nonlinear circuit operating under low-distortion conditions is given by:

$$v_{in}(t) = \exp(j\omega_o t) \tag{2.4.a}$$

the output signal is of the form

$$v_{out}(t) = \sum_k A_k \cdot \exp(jk\omega_o t) \tag{2.4.b}$$

Note that the input voltage of (2.4.a) is complex; it can never be applied in reality but is treated here in a mathematical way. The coefficients A_k in (2.4.b) are complex numbers, functions of ω_o, that can be readily found from network analysis. Some examples will be discussed in the next paragraph.

These coefficients A_k contain enough information to calculate the harmonic distortion when a realistic input signal is applied [4]: when the input signal is of the form of expression (2.1.a):

$$v_{in}(t) = V_{in} \cdot \cos(\omega_o t) \tag{2.1.a}$$

then is the output signal of the form of (2.1.b):

$$v_{out}(t) = \sum_k V_k \cdot \cos(k\omega_o t + \varphi_k) \tag{2.1.b}$$

where the parameters V_k and φ_k can be calculate from the coefficients A_k in expression (2.4.b):

$$V_k = \frac{|A_k|}{2^{k-1}} \cdot V_{in}^k \quad \text{and} \quad \varphi_k = \arg(A_k) \tag{2.5.a}$$

and according to (2.2):
$$HD_k = \frac{1}{2^{k-1}} \cdot \frac{|A_k|}{|A_1|} \cdot V_{in}^{k-1} \tag{2.5.b}$$

A mathematical proof of the expressions (2.5.a) is given in appendix 2.A. From (2.5.b), it can be seen that the second harmonic distortion is proportional to the input amplitude, the third harmonic distortion is proportional to the amplitude square, etc...

The commonly used distortion calculation technique for **resistive circuits**, starting from the nonlinear DC-transfer characteristic [1], is a special case of the calculation technique presented above, only valid for low-frequency quasistatic input signals. This calculation technique for quasistatic signals proceeds as follows:

Consider a nonlinear resistive circuit, characterised by its nonlinear input-output DC-transfer characteristic:

$$v_{out} = \sum_k a_k \cdot v_{in}{}^k \tag{2.6}$$

When now a (mathematical) low-frequency input signal of the form of (2.4.a) is applied to this circuit, the output voltage will be given by:

$$v_{out}(t) = \sum_k a_k \cdot [\exp(j\omega_o t)]^k \tag{2.7.a}$$

$$= \sum_k a_k \cdot \exp(jk\omega_o t) \tag{2.7.b}$$

This expression is of the same form as (2.4.b). Hence, for small frequencies ω_o, the coefficients A_k in (2.4.b) are nothing else than the coefficients a_k of the nonlinear DC-transfer characteristic (2.6). As mentioned above, the coefficients A_k of (2.4.b) (and thus the coefficients a_k of expression (2.6)) contain sufficient information to calculate the harmonic distortion for a low-frequency sinusoidal input signal:

$$v_{out}(t) = \sum_k V_k \cdot \cos(k\omega_o t + \varphi_k) \tag{2.1.b}$$

where $\qquad V_k = \dfrac{a_k}{2^{k-1}} \cdot V_{in}{}^k \quad$ and $\varphi_k = 0 \tag{2.8}$

and hence: $\qquad HD_k = \dfrac{1}{2^{k-1}} \cdot \dfrac{a_k}{a_1} \cdot V_{in}{}^{k-1} \tag{2.9}$

This expression is equivalent to expression (2.4.b). It is obvious that a calculation based upon the nonlinear DC-transfer characteristic yields no frequency or phase information: expression (2.9) is independent of ω_o.

2.2.c. Basic distortion expressions

In this paragraph, distortion expressions for the basic transistor configurations are summarised, so that the reader can refer to them in next sections.

2.2.c.1: A common-emitter stage

The common-emitter amplifier stage is a first example of a nonlinear dynamic circuit. With this example, we will show here that it is often relatively simplt to calculate the coefficients A_k of expressions (2.4.b).

When the circuit of Fig. 2.1.a. is driven with a voltage given by

$$v_{in}(t) \; = \; \exp(j\omega_o t) \tag{2.4.a}$$

and when a simple square-law transistor model is assumed, the transistor drain current equals:

$$i_D(t) \; = \; K_p \cdot W/L \cdot (V_{GS} + \exp(j\omega_o t) - V_T)^2 \tag{2.10.a}$$

$$= \; I_Q + g_m \cdot \exp(j\omega_o t) + \frac{g_m}{2(V_{GS}-V_T)} \cdot \exp(j2\omega_o t) \tag{2.10.b}$$

where I_Q the quiescent drain current and V_{GS} is the DC gate bias voltage.
According to expression (2.4.b), The output voltage is of the form:

$$v_{out}(t) \; = \; V_D + A_1 \cdot \exp(j\omega_o t) + A_2 \cdot \exp(j2\omega_o t) \tag{2.11}$$

and thus: $i_D(t) = I_Q - g_L \cdot v_{out}(t) - C_L \cdot dv_{out}/dt \tag{2.12}$

$$= \; I_Q - A_1 \cdot (g_L + j\omega_o C_L) \cdot \exp(j\omega_o t)$$

$$- \; A_2 \cdot (g_L + 2j\omega_o C_L) \cdot \exp(2j\omega_o t) \tag{2.13}$$

Equating the terms with same frequency in (2.10.b) and (2.13) yields:

$$A_1 = g_m \cdot Z_L(\omega_o) \qquad \text{and} \quad A_2 = \frac{1}{2(V_{GS}-V_T)} \cdot g_m \cdot Z_L(2\omega_o) \tag{2.14}$$

$$\text{with } Z_L(\omega) = 1/(g_L + j\omega C_L) \tag{2.15}$$

and according to (2.5.b):

$$HD_2 = \frac{1}{4} \frac{Z_L(2\omega_o)}{Z_L(\omega_o)} \cdot \frac{V_{in}}{V_{GS}-V_T} \tag{2.16}$$

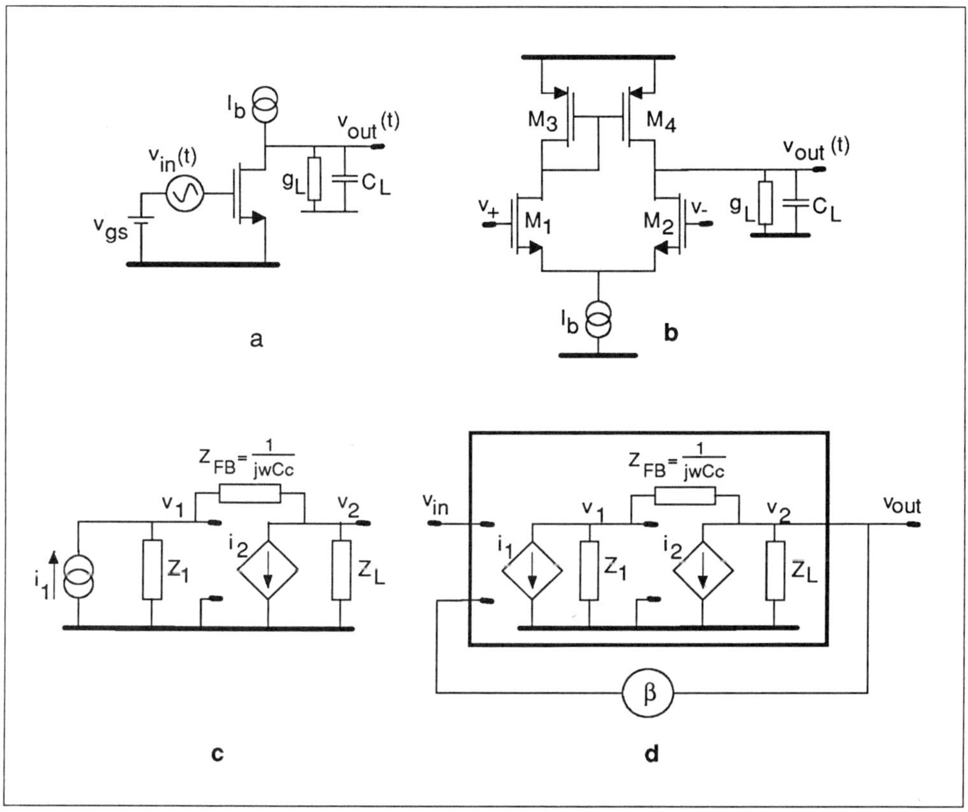

Fig. 2.1. Basic Transistor configurations
  a) common emitter stage,
  b) differential pair,
  c) Miller compensated amplifier,
  d) Miller compensated amplifier with feedback

For frequencies, well below the pole frequency, this result agrees with the result obtained with the Taylor series method [1]. But for high frequencies, the magnitude of $Z_L(2\omega_o)$ is only half of the magnitude of $Z_L(\omega_o)$ and hence, there is a difference of 6 dB.

Short-channel MOS transistors exhibit a more linear I/V relationship than long-channel devices [5] and the distortion will be less. In this text however, (2.16) will be used as a safe upper limit, even for short-channel devices.

In the calculations presented above, a square-law MOS I/V characteristic is assumed. As a result, the common-emitter stage generates no third- or higher harmonic distortion. In reality, these harmonics will occur. However, it is impossible to estimate them with a simple square-law assumption.

2.2.c.2. A differential pair

When the differential pair of Fig. 2.1.b. is perfectly symmetrical, with an ideal bias current source, a distortion calculation similar to the previous section yields:

$$HD_2 = 0 \text{ and } HD_3 = \frac{1}{64} \cdot \frac{Z_L(3\omega_o)}{Z_L(\omega_o)} \cdot \frac{V_{in}^2}{(V_{GS}-V_T)^2} \tag{2.17}$$

2.2.c.3 A Miller compensated output stage

In Fig. 2.1.c., the principle small-signal schematic of a Miller-compensated amplifier is depicted. The compensation capacitor forms a feedback path around the output stage. The objective of this paragraph is to investigate the influence of this internal feedback to the open loop distortion of the amplifier [6].

The second stage of this amplifier is modelled as a nonlinear memoryless transconductance:

$$i_2(t) = gm_{21} \cdot v_1(t) + gm_{22} \cdot v_2(t)^2 + gm_{23} \cdot v_1(t)^3 + \ldots \tag{2.18}$$

The coefficients gm_{21}, gm_{22} and gm_{23} can be calculated from a Taylor expansion of the nonlinear transconductance around the operating point. Since only the nonlinearity of the output stage is considered., the input stage is modelled as a linear current source. The impedances Z_1, Z_{FB} and Z_L are assumed linear and frequency dependent. The distortion of this circuit can be calculated using the method described above: Assume $i_1(t)$ is given by:

$$i_1(t) = \exp(j\omega_o t) \tag{2.19}$$

The voltages on node 1 and 2 can generally be expressed as a sum of exponentials:

$$v_1(t) = A_{11} \cdot \exp(j\omega_o t) + A_{12} \cdot \exp(2j\omega_o t) + A_{13} \cdot \exp(3j\omega_o t) + \ldots \tag{2.20}$$

$$v_2(t) = A_{21} \cdot \exp(j\omega_o t) + A_{22} \cdot \exp(2j\omega_o t) + A_{23} \cdot \exp(3j\omega_o t) + \ldots \tag{2.21}$$

Expressing Kirchoff's law on nodes 1 and 2 and equating the term with same frequency yields:

$$A_{21} = \frac{-1}{j\omega_o C_c \cdot [1+1/T(\omega_o)]} \tag{2.22}$$

$$A_{22} = \frac{gm_{22}}{gm_{21}} \cdot \frac{G(2\omega_o)}{1+T(2\omega_o)} \cdot \frac{A_{21}^2}{G(\omega_o)^2} \tag{2.23}$$

$$\text{and} \quad A_{23} = [-2 \cdot (\frac{gm_{22}}{gm_{21}})^2 \cdot \frac{T(2\omega_o)}{1+T(2\omega_o)} + \frac{gm_{23}}{gm_{21}} \cdot \frac{G(3\omega_o)}{1+T(3\omega_o)}] \cdot \frac{A_{21}^3}{G(\omega_o)^3} \tag{2.24}$$

$$\text{with} \quad G(\omega) = gm_{21} \cdot [Z_L(\omega) // (Z_1(\omega) + Z_{FB}(\omega))] \tag{2.25}$$

$$\text{and} \quad T(\omega) = G(\omega) \cdot \frac{Z_1(\omega)}{Z_1(\omega) + Z_{FB}(\omega)} \tag{2.26}$$

Note that G(ω) is the open-loop gain of the amplifier stage and T(ω) is the loop gain around this stage.

The coefficient A_{21} to A_{23} contain sufficient information to calculate the nonlinear circuit behaviour for realistic input signals: when the circuit is driven with a current equal to

$$i_1(t) = I_1 \cdot \cos(\omega_o t)$$

the harmonic distortion components are given by:

$$HD_{2f} = \frac{1}{2} \frac{A_{22}}{A_{21}} I_{in} = \frac{1}{2} \frac{gm_{22}}{gm_{21}} \cdot \frac{G(2\omega_o)}{G(\omega_o)} \cdot \frac{V_{out}}{G(\omega_o)} \cdot \frac{1}{1+T(2\omega_o)} \tag{2.27}$$

$$HD_{3f} = \frac{1}{4} [-2 (\frac{gm_{22}}{gm_{21}})^2 \cdot \frac{T(2\omega_o)}{1+T(2\omega_o)} + \frac{gm_{23}}{gm_{21}}] \cdot \frac{G(3\omega_o)}{G(\omega_o)} \cdot \frac{V_{out}^2}{G(\omega_o)^2} \cdot \frac{1}{1+T(3\omega_o)} \tag{2.28}$$

where V_{out} is the output amplitude, equal to $A_{21}.I_1$

In order to examine the influence of the feedback capacitance, these distortion terms have to be compared with the distortion without internal feedback, which can be found by setting T(ω) equal to zero in (2.27) and (2.28):

$$HD_{2o} = \frac{1}{2} \cdot \frac{gm_{22}}{gm_{21}} \cdot \frac{G(2\omega_o)}{G(\omega_o)} \cdot \frac{V_{out}}{G(\omega_o)} \tag{2.29}$$

$$HD_{3o} = \frac{1}{4} \cdot \frac{gm_{23}}{gm_{21}} \cdot \frac{G(3\omega_o)}{G(\omega_o)} \cdot \frac{V_{out}^2}{G(\omega_o)^2} \tag{2.30}$$

When the second and third open loop harmonic are assumed of about the same order of magnitude and much smaller than the fundamental signal,

$$gm_{22}^2 \text{ is much smaller than } gm_{21} \cdot gm_{23} \qquad (2.31)$$

and the harmonic distortion components with feedback are given by:

$$HD_{2f} = \frac{HD_{2o}}{1+T(2\omega_o)} \qquad (2.32)$$

$$HD_{3f} = \frac{HD_{3o}}{1+T(3\omega_o)} \qquad (2.33)$$

Compared with the open loop distortion, the second and third harmonic distortion are a factor $[1+T(2\omega_o)]$ resp. $[1+T(3\omega_o)]$ smaller for the same output voltage. This result can be generalised: feedback divides the distortion by the gain around the feedback loop, **measured at the considered harmonic frequency.**

For the Miller compensated amplifier, this is a convenient way to suppress distortion without requiring a too excessive external loop gain: for the circuit of Fig. 2.1.d., the distortion from the output stage will be divided by the product of two loop gains:

$$HD_{2f} = HD_{2o}/\left[\frac{\beta \cdot 2\pi \cdot GBW}{2\omega_o} \cdot \frac{2\pi \cdot p_2}{2\omega_o}\right] \qquad (2.34)$$

where p_2 is the second pole of the amplifier. Although the condition (2.31) is questionable for the inner loop, it certainly holds for the outer feedback loop. Note that the distortion suppression due to the inner loop is always larger than the one due to the outer loop.

With a similar calculation, it can be shown that also the distortion due to a nonlinear load (nonlinear Z_L in Fig. 2.1.c.) or a nonlinear internal impedance (nonlinear Z_1 in Fig. 2.1.c.) is suppressed by the product of two loop gains. Distortion due to a nonlinear input stage transconductance is only suppressed by the outer feedback loop and the nonlinearities from the feedback factor β are not suppressed at all.

Note that for frequencies well above the dominant pole, the phase of $T(\omega_o)$ and $G(\omega_o)$ (see (2.25) and (2.26)) equals -90 ^{o}. Hence, the coefficient of V_{out}^2 in (2.28) has a phase of 270°. From expressions (2.17) and (2.28), it can be seen that third harmonic distortion from the second stage can never compensate for the distortion of the input stage because there is a 90 degrees phase difference.

2.3. THE RELATIONSHIP BETWEEN THE CMRR AND THE HARMONIC DISTORTION OF A DIFFERENTIAL INPUT AMPLIFIER

According to basic theories (see expression 2.17), a differential input pair only produces odd harmonics which can be suppressed indefinitely by applying more and more negative feedback. Due to mismatches between the input transistors, even harmonics will be produced as well [7]. Although this effect is well known among circuit engineers, it is often treated as only a second-order effect. However, distortion measurements on a test amplifier presented below show that the second harmonic distortion is normally even larger than the third one. Therefore, the effects of mismatches are of prime importance for distortion analysis.

The common-mode gain is an other effect which is often neglected in distortion analysis. When using a differential pair as a non-inverting buffer, the common-mode input voltage is much larger than the differential input voltage and is not affected by the negative feedback. Moreover, the nonlinearity of the common-mode gain causes additional distortion which is not suppressed by the negative feedback and therefore becomes the major distortion contribution.

As is shown below, for large feedback factors, the differential input voltage of a non-inverting amplifier is determined by the finite common-mode rejection ratio (CMRR) rather than by the differential gain. This is a third effect which is often neglected in distortion analyses. This CMRR limits the amount of feedback which can usefully be applied to suppress the distortion.

In the analysis below, the differential pair is treated as a nonlinear circuit with two independent inputs (the differential input voltage and the common-mode input voltage). Since it is only important to give a qualitative explanation of the effects mentioned above, only the second harmonic distortion is calculated. An analysis of the third harmonic is quite involved and is omitted.

2.3.a. Expressions for the second harmonic distortion of a differential pair with mismatches

In Fig. 2.2.a. , a differential pair is depicted. Assume the I/V relations of the elements are given by:

$$i_1(t) = a_1 \cdot v_{gs1}(t) + a_2 \cdot v_{gs1}(t)^2 + \ldots \qquad (2.35a)$$

$$i_2(t) = b_1 \cdot v_{gs2}(t) + b_2 \cdot v_{gs2}(t)^2 + \ldots \qquad (2.35b)$$

$$i_s(t) = (g_1 + g_2 \cdot v_s + \ldots) v_s + (C_1 + C_2 \cdot v_s + \ldots) dv_s/dt \qquad (2.35c)$$

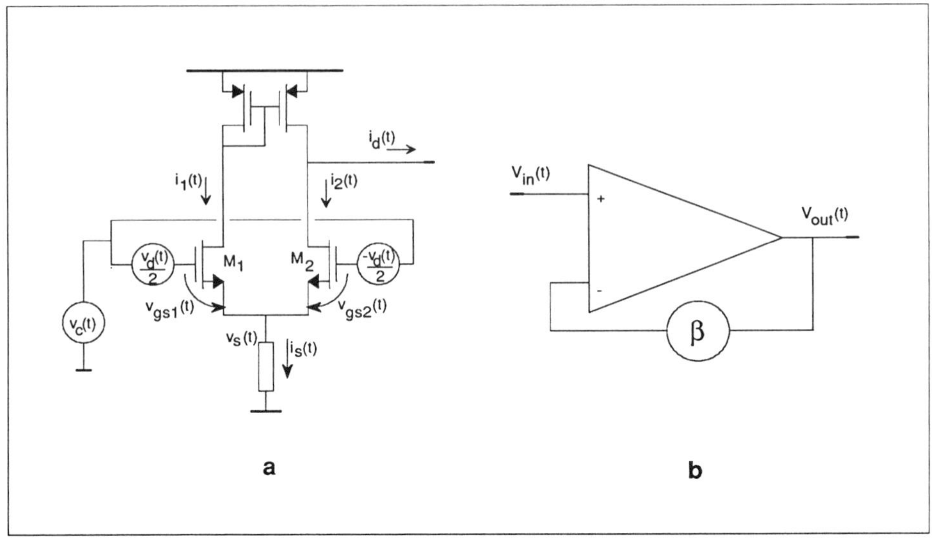

Fig. 2.2.   a) an assymetrical differential pair
b) a non-inverting amplifier

where i_1, i_2, i_s, v_{gs1}, v_{gs2} and v_s are the ac values.   The parameters a_1 and b_1 represent the transconductances of M_1 and M_2 while a_2 and b_2 represent their second-order nonlinearities.   Due to mismatches, a_1 and a_2 can be different from b_1 and b_2. These parameters depend on the operating point [5] [8] [9].   In order to simplify the problem, no bulk effect is taken into account.   The parameters g_1 and g_2 represent the output conductance of the current source and its second-order nonlinearity while C_1 and C_2 represent the nonlinear capacitance on the common-source node.   Note that the distortion caused by the current mirror or by the nonlinear load is not considered.

The circuit of Fig. 2.2.a. is driven simultaneously with a differential-mode and a common-mode signal:

$$v_d(t) \; = \; V_d.\exp(j\omega_d t) \tag{2.36a}$$

$$v_c(t) \; = \; V_c.\exp(j\omega_c t) \tag{2.36b}$$

The differential output current $i_d = i_1 - i_2$ can be calculated from (2.35) and (2.36). This yields:

$$i_d(t) \; = \; gm_{d1}.V_d.\exp(j\omega_d t) \; + \; gm_{d2}.V_d^2.\exp(2j\omega_d t)$$

$$+ \; gm_{c1}(\omega_c).V_c.\exp(j\omega_o t) \; + \; gm_{c2}(\omega_c).V_c^2.\exp(2j\omega_o t)$$

$$+ \; gm_{cd}(\omega_c).V_c.V_d.\exp(j(\omega_c + \omega_d)t) \; + \; ... \tag{2.37}$$

in which

$$gm_{d1} = (a_1 + b_1)/2$$

$$gm_{c1}(\omega_c) = (1-b_1/a_1)/2 \cdot y_1(\omega_c) \qquad \text{with} \qquad y_1(\omega) = g_1 + j\omega C_1$$

$$gm_{d2} = [a_2 \cdot (b_1/a_1)^3 - b_2]/4$$

$$gm_{c2}(\omega_c) = (1-b_1/a_1)/2 \cdot y_2(\omega_c) \qquad \text{with} \qquad y_2(\omega_o) = g_2 + j\omega C_2$$

$$gm_{cd}(\omega_c) = y_1(\omega_c) \cdot a_2/a_1 \tag{2.38}$$

The physical meaning of these coefficients is best explained by stating that the total differential output current (2.37) consist of three terms: the first term is caused by the differential input only: the coefficient gm_{d1} represent the linear behaviour of the transconductance while gm_{d2} represents the second-order nonlinearity. Note that in the ideal case without mismatches, gm_{d2} equals zero. The second term of (2.37) is caused by the common-mode input voltage. The meaning of gm_{c1} and gm_{c2} is similar to that of gm_{d1} and gm_{d2} but now for the common-mode input voltage. The third term, caused by the interaction of both inputs, is typical for a nonlinear circuit. In a linear circuit, this interaction is impossible, according to the superposition principle. The coefficients gm_{cd} and gm_{c2} depend upon the common-source conductance and are therefore much smaller than gm_{d2}. However, since for a non-inverting amplifier as depicted in Fig. 2.2.b., the common-mode input voltage is much larger than the differential input voltage, their contribution to the second harmonic is significant.

2.3.b. The second harmonic distortion of a non-inverting amplifier

When an amplifier is used as a non-inverting buffer, the common-mode input voltage approximately equals the amplifier input voltage v_{in} and is almost not affected by the loop gain. Therefore, it is interesting to investigate how the nonlinearities of (2.37) are suppressed by the negative feedback. Under the assumptions that $T(\omega_o)$ is much larger than unity and β is frequency independent, a calculation of the second harmonic distortion for the circuit of Fig. 2.2.b. yields:

$$HD_2 = \frac{1}{2} \left\{ \frac{gm_{c2}(\omega_o)}{gm_{d1}} + \frac{gm_{cd}(\omega_o)}{gm_{d1}} \cdot \frac{1-T(\omega_o)/CMRR(\omega_o)}{1+T(\omega_o)} + \right.$$

$$\left. \frac{gm_{d2}(\omega_o)}{gm_{d1}} \cdot \left[\frac{1-T(\omega_o)/CMRR(\omega_o)}{1+T(\omega_o)} \right]^2 \right\} \cdot V_i \tag{2.39}$$

with V_i the input amplitude and ω_o the input signal frequency. $T(\omega_o)$ is the loop gain and $CMRR(\omega_o)$ is the common-mode rejection ratio, given by

$$CMRR(\omega_o) = gm_{d1}/gm_{c1}(\omega_o) \tag{2.40}$$

For sake of illustration, one can evaluate (2.39) assuming a unity gain buffer with a gain-bandwidth of 100 kHz, a load capacitance of 10 pF and a CMRR of 80 dB, driven by a 500 mV input voltage at 1 kHz. From this, one finds $T(\omega_o)=100$, $a_1=gm_1=2\pi.C_L.GBW=6.3E\text{-}6$ A/V. Furthermore, according to (2.10), $a_2=g_{m1}/2(V_{GS}\text{-}V_T) = 1.6E\text{-}5$ A/V^2 for a $(V_{GS}\text{-}V_T)$ of 0.2V. Suppose finally that $a_1\text{-}b_1=1\%.a_1$ and $a_2\text{-}b_2=1\%.a_2$. Using expressions (2.39) and (2.40), and assuming that y_1 varies with 50% per volt v_s, the terms of (2.39) yield in order -98 dB, -73 dB, -126 dB. Clearly, the largest distortion contribution is caused by the nonlinear common-mode gain, even for amplifiers with a rather large CMRR.

In order to obtain some insight in expression (2.39), two cases can be distinguished:

Case one: $T(\omega_o) << CMRR(\omega_o)$

In this case, expression (2.39) reduces to:

$$HD_2 = \frac{1}{2} \left\{ \frac{gm_{c2}(\omega_o)}{gm_{d1}} + \frac{gm_{cd}(\omega_o)}{gm_{d1}} \cdot \frac{1}{1+T(\omega_o)} + \frac{gm_{d2}}{gm_{d1}} \cdot \frac{1}{[1+T(\omega_o)]^2} \right\} . V_i$$

$$(2.41)$$

For a constant input voltage, the contribution of gm_{d2} to HD_2 is suppressed by the feedback with a factor $[1+T(\omega_o)]^2$, as expected. However, the contribution of gm_{cd} is only suppressed with a factor $[1+T(\omega_o)]$ while the contribution of gm_{c2} is not suppressed at all. Especially when the loop gain is large, the common-mode gain nonlinearities are more important than the differential gain nonlinearity.

For a unity gain amplifier with an input voltage of several volts, and when the nonlinearity of $y_1(\omega_o)$ is in the order of 20% to 50% per volt v_s, the first term of (2.41) is of the same order of magnitude as the CMRR.

Case two: $T(\omega_o) >> CMRR(\omega_o)$

In this case (2.39) reduces to:

$$HD_2 = \frac{1}{2} \left\{ \frac{gm_{c2}(\omega_o)}{gm_{d1}} - \frac{gm_{cd}(\omega_o)}{gm_{d1}} \cdot \frac{1}{CMRR(\omega_o)} + \frac{gm_{d2}}{gm_{d1}} \cdot \frac{1}{CMRR(\omega_o)^2} \right\} . V_i$$

$$(2.42)$$

Comparing (2.42) with (2.41), one can conclude that, when the loop gain increases above the CMRR, no further distortion improvement is achieved. The distortion becomes independent of the loop gain. This can be explained as follows: Due to the finite CMRR, a common-mode input voltage produces an output signal. The negative feedback then will produce a differential voltage between the inputs to compensate for the common-mode input voltage. When the CMRR is smaller than the loop gain, the differential input voltage is determined by

$$\frac{V_{common,\,in}}{\mathrm{CMRR}(\omega_O)} = \frac{\beta(\omega_O)\cdot V_{out}}{\mathrm{CMRR}(\omega_O)} \quad \text{rather than by} \quad \frac{V_{out}}{\mathrm{Gain}(\omega_O)}$$

in which $\beta(\omega_O)$ denotes the feedback factor. The negative feedback can no longer influence the differential input voltage and therefore has no influence on the distortion.

2.3.c. Distortion measurements on a test amplifier

In order to illustrate the importance of the mismatches and of the common-mode gain, distortion measurements are performed on the circuit of Fig. 2.3., which is a Miller compensated operational amplifier. The load capacitance at the output is small (about 10 pF), so that the second pole is large and the second stage provides only a minor contribution to the distortion. (see equations 2.27 to 2.33).

In Fig. 2.4, the measured second and third harmonic distortion of the amplifier when used as an inverting buffer (See Fig. 2.3.b.) are presented. For frequencies below 10 kHz, both the second and third harmonic distortion are below noise level. But for frequencies where the distortion becomes measurable, the second harmonic is larger than the third one. Since HD_3 is proportional to ω^3 while HD_2 is only proportional to ω^2, the difference is even more pronounced at lower frequencies.

It could be expected that the second harmonic is larger than the third one: the second harmonic distortion is proportional to the differential input amplitude while the third one is proportional to the amplitude square. Since for operational amplifiers the differential input amplitude is usually very small, even a small mismatch is sufficient to cause a second harmonic distortion, larger than the third one. This illustrates the importance of the mismatches.

Fig. 2.4. also compares the distortion of the amplifier when used as a non-inverting buffer (see Fig. 2.3.c.) with the distortion when used as an inverting amplifier (see Fig. 2.3.b.). In the two configurations, the load is the same. The second harmonic distortion for the non-inverting buffer is over 20 dB larger for frequencies up to 10 kHz. This increase can only be caused by the input common-mode voltage since this is the only difference between the two circuits. This illustrates the importance of the common-mode gain.

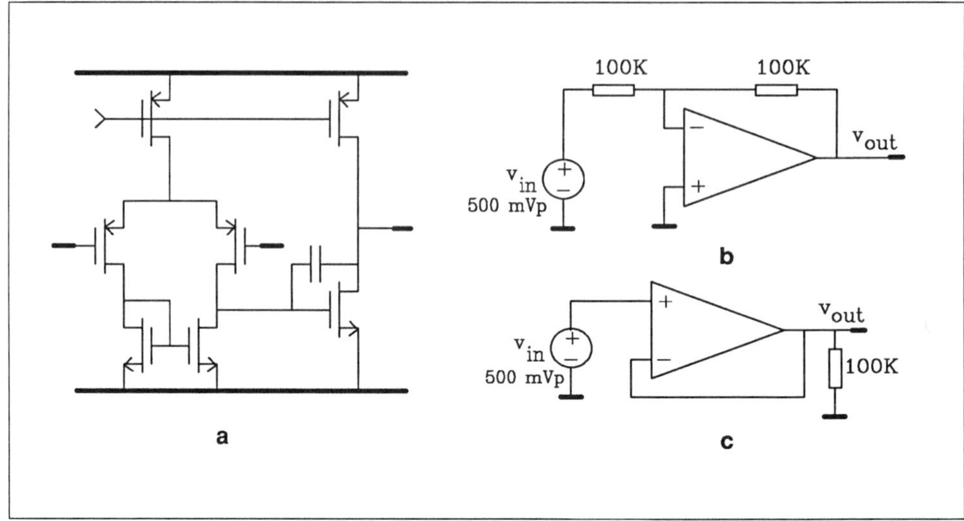

Fig. 2.3. The test amplifier:
  a) schematic
  b) used as an inverting amplifier
  c) used as a non-inverting amplifier

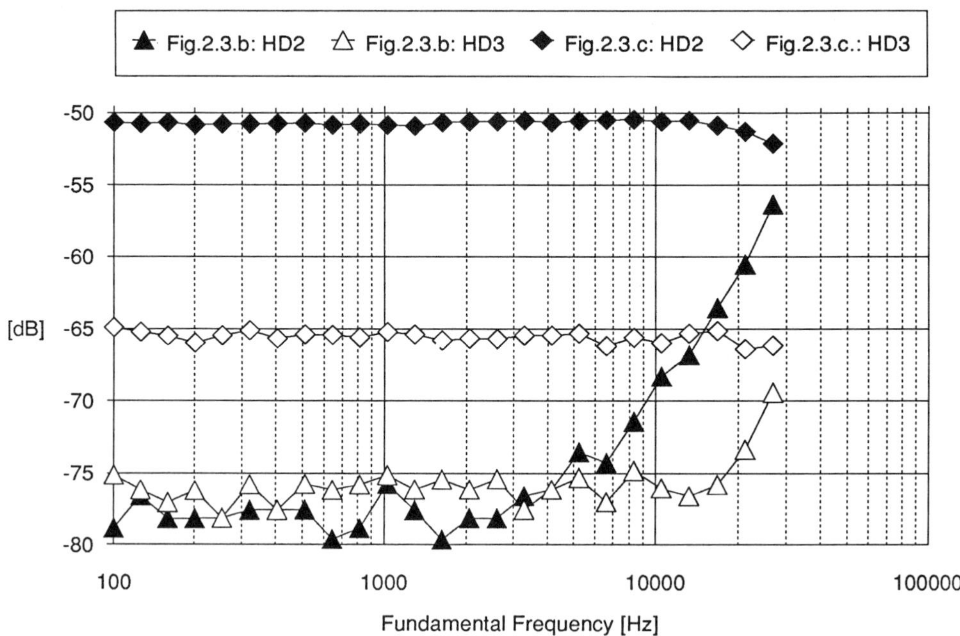

Fig. 2.4. Distortion measurement results for the circuits of Fig. 2.3.

2.4. THE SECOND HARMONIC DISTORTION OF A CLASS A AMPLIFIER WITH LIMITED POWER SUPPLY REJECTION RATIO

In Fig. 2.5., an amplifier with a load conductance $Y_L(\omega)$ is depicted. Due to the impedance $Z_S(\omega)$ in the negative supply line, the supply voltage becomes variable when the output current is drawn from this supply line. In a similar way as the CMRR of a non-inverting buffer, the power supply rejection ratio (PSRR) will influence the harmonic distortion [7]. In order to analyse this effect, we will drop the influence of the finite CMRR and focus on the PSRR for the negative power supply.

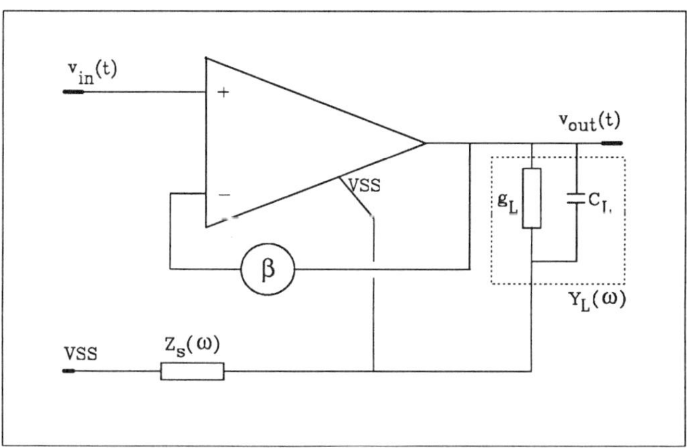

*Fig. 2.5. An amplifier with load conductance and Supply
line impedance*

2.4.a. Expressions for the second harmonic distortion for a Class A amplifier with finite PSRR

In the same way as in equation (2.37), the output voltage of the circuit of Fig. 2.5 can be expressed as:

$$v_{out}(t) = [A_{d1}(\omega_o) \cdot V_d + A_{s1}(\omega_o) \cdot V_s] \cdot \exp(j\omega_o t) +$$
$$[A_{d2}(\omega_o) \cdot V_d^2 + A_{ds}(\omega_o t) \cdot V_d \cdot V_s + A_{s2}(\omega_o) \cdot V_s^2] \cdot \exp(2j\omega_o t)$$
$$+ \ldots \tag{2.43.a}$$

with $\quad V_s = V_{out} \cdot Y_L(\omega_o) \cdot Z_S(\omega_o) \tag{2.43.b}$

In these expressions, A_{d1} is the differential gain, A_{s1} is the power-supply gain, A_{d2} is the differential gain nonlinearity, A_{s2} is the power-supply gain nonlinearity and A_{ds} is the cross-term. V_d is the differential input amplitude and V_s is the amplitude of the power-supply variations.

The closed loop second harmonic distortion of this circuit can be expressed with an equation similar to (2.39). The feedback factor from output to input is denoted as "$\beta(\omega)$" and the gain from the output to the power-supply line ($Y_L(\omega).Z_S(\omega)$ in Fig. 2.5.) by "$\gamma(\omega)$". Again, two cases can be distinguished:

Case 1: $T(\omega_o) << PSRR(\omega_o).\beta(\omega_o)/\gamma(\omega_o)$

In this case, the second harmonic distortion is given by:

$$HD_2 = \frac{1}{2} \cdot \frac{1+T(\omega_o)}{1+T(2\omega_o)} \cdot \left\{ \frac{A_{d2}(\omega_o)}{A_{d1}(\omega_o)} \cdot \frac{1}{[1+T(\omega_o)]^2} + \frac{A_{ds}(\omega_o)}{A_{d1}(\omega_o)} \cdot \frac{\gamma(\omega_o)/\beta(\omega_o)}{1+T(\omega_o)} + \right.$$

$$\left. + \frac{A_{s2}(\omega_o)}{A_{d1}(\omega_o)} \cdot [\gamma(\omega_o)/\beta(\omega_o)]^2 \right\} \cdot V_i \qquad (2.44)$$

As can be seen, the term due to A_{ds} is only suppressed by a factor $[1+T(\omega_o)]$ and the term from A_{s2} is not suppressed at all.

Case 2: $T(\omega_o) >> PSRR(\omega_o).\beta(\omega_o)/\gamma(\omega_o)$

In this case, the second harmonic distortion calculations yield:

$$HD_2 = \frac{1}{2} \cdot \frac{1+T(\omega_o)}{1+T(2\omega_o)} \cdot \left[\frac{\gamma(\omega_o)}{\beta(\omega_o)} \right]^2 \cdot \left\{ \frac{A_{d2}(\omega_o)}{A_{d1}(\omega_o)} \cdot \frac{1}{PSRR(\omega_o)^2} + \right.$$

$$\left. - \frac{A_{ds}(\omega_o)}{A_{d1}(\omega_o)} \cdot \frac{1}{PSRR(\omega_o)} + \frac{A_{s2}(\omega_o)}{A_{d1}(\omega_o)} \right\} \cdot V_i \qquad (2.45)$$

This equation is similar to expression (2.42): When the PSRR becomes too small, the distortion becomes independent of the loop gain. The explanation is also similar: the negative feedback produces a differential voltage between the input nodes in order to compensate for the supply voltage variations. When the influence of the supply voltage variations is large, the differential input voltage is determined mainly by the supply voltage variations and becomes independent of the loop gain.

2.4.b. Distortion measurement results on a test amplifier

In order to illustrate the effect of the power supply gain, distortion measurements are performed on the amplifier of Fig. 2.3.a., driving a heavy load and with a large power supply impedance (see Fig. 2.6.a.). The results are presented in Fig. 2.6.b. and compared with the harmonic distortion of the inverting amplifier of Fig. 2.3.b.

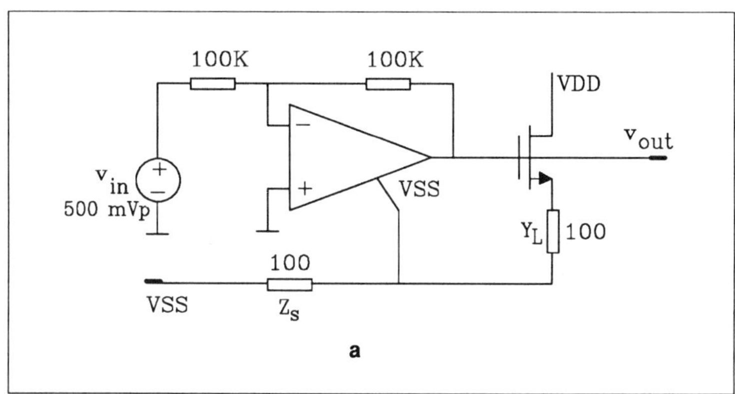

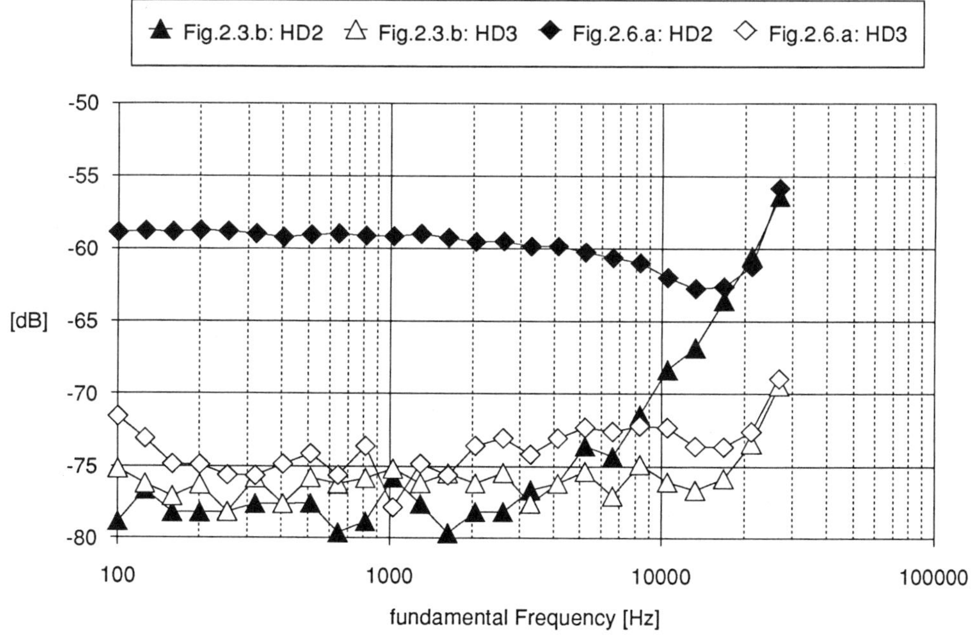

Fig. 2.6.   *a) The test amplifier with a heavy load and a Supply impedance*
b) The measured harmonic distortion

Since the input impedance of the source follower in Fig. 2.6.a. is very high, both circuits have the same load. For frequencies up to 10 kHz, the second harmonic distortion is over 15 dB worse for the circuit of Fig. 2.3.b. This is due to the varying power supply voltage, since this is the only difference between the two circuits. This illustrates the importance of the power supply gain.

For the circuit of Fig. 2.3.a., the current through the positive supply line is constant. Therefore, the positive power supply gain has no influence on the second harmonic distortion.

2.5. DISTORTION DUE TO THERMAL FEEDBACK

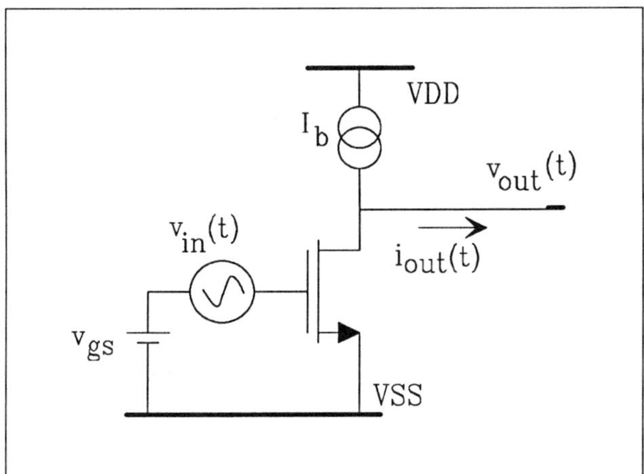

Fig. 2.7. A Class A amplifier

When an amplifier drives current to its load, the power dissipation in the output stage is a square law function of the output voltage. Since this power is dissipated on chip, it causes temperature variations in the input transistors, resulting in variations of their electrical characteristics. Due to mismatches or due to an asymmetrical layout, this results in offset voltage variations. This process is well known as thermal feedback [10]. Since thermal feedback is signal dependent and inherently nonlinear, it causes distortion. The purpose of this section is to investigate the three components of this distortion generation mechanism - thermal power generation, temperature variation, offset voltage variation - more in detail and estimate the harmonic distortion.

2.5.a. The power generation

For the Class A amplifier of Fig. 2.7., the on chip power dissipation as a function of time equals:

$$p(t) = I_b \cdot (V_{DD} - V_{SS}) + i_{out}(t) \cdot V_{SS} - i_{out}(t) \cdot v_{out}(t) \tag{2.46}$$

where V_{SS} is negative.

When the output voltage is a sinusoidal signal, p(t) varies with time and the last term of (2.46) causes a contribution at double frequency, with amplitude:

$$P(2\omega_o) = \frac{1}{2} \cdot V_{out}^2 \cdot |Y_L(\omega_o)| \tag{2.47}$$

in which V_{out} is the output amplitude and Y_L is the (complex) load conductance.

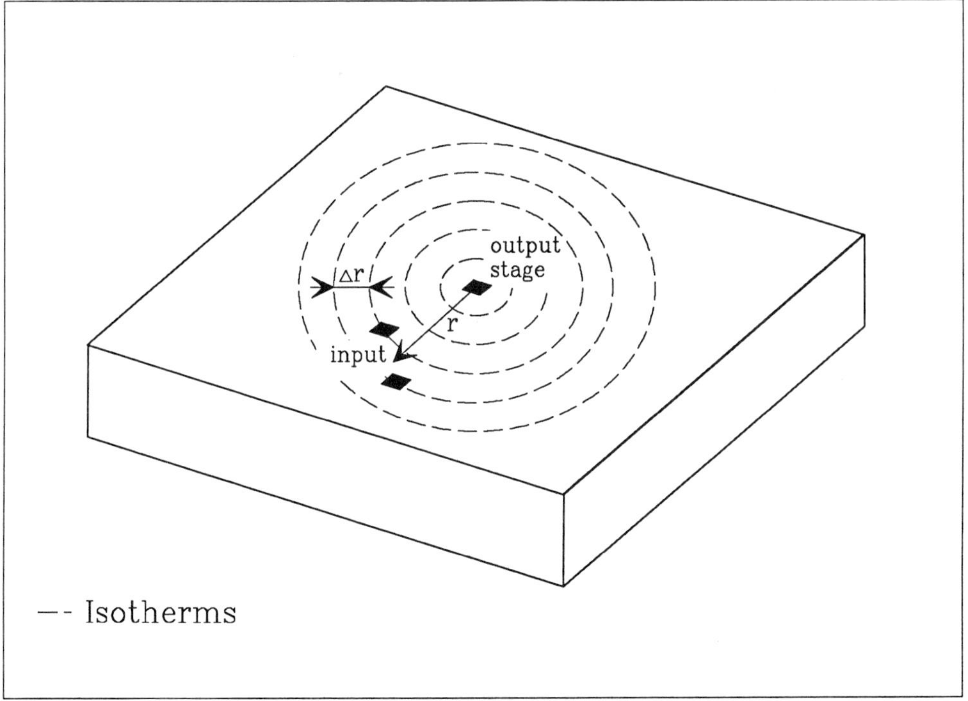

*Fig. 2.8. An illustration of the thermal feedback
mechanism.*

2.5.b. The temperature variation

The thermal power generated by the output stage is spread out over the chip by thermal conduction and finally dissipated by the package. Since the differential equations governing this process are linear, they can be solved in the frequency domain: assume that the power generated at double signal frequency is of the form:

$$p(t) = P(2\omega_o) \cdot \exp(2j\omega_o t) \qquad (2.48)$$

If we assume that this power is generated in an infinitely small chip area as depicted in Fig. 2.8, the temperature distribution over the chip can be readily calculated:

$$T(r,t) = \frac{P(2\omega_o)}{2\pi\lambda} \cdot \frac{\exp[-\sqrt{\omega_o/\delta}\ r]}{r} \cdot \exp[j2\omega_o(t - \frac{r}{\sqrt{4\omega_o\delta}})] \qquad (2.49)$$

with r the distance from the power source, λ the thermal conductivity of silicon (1.5 W/cmK) and δ the thermal diffusivity (0.9 cm^2/s). Expression (2.49) is valid for distances smaller than the wafer thickness.

For example, an amplifier driving a 100 Ohm load with an output amplitude of 2 Volt at 50 Hz causes an on chip power dissipation at double frequency with 20 mW amplitude. The temperature wave at 100μm distance has an amplitude of 0.16 Kelvin. At 1000μm, the temperature variation is only 0.001 K. Note that the temperature amplitude is larger for lower signal frequencies.

2.5.c. The offset voltage variation

When the two amplifier input transistors are placed on a different distance to the thermal power source as depicted in Fig. 2.8., the temperature difference has an amplitude

$$|\Delta T(2\omega_o)| = T(r+\Delta r,t) - T(r,t)$$

$$= |T(r,t)| \cdot |\frac{1}{r} + (1+j) \cdot \sqrt{\frac{\omega_o}{\delta}}| \cdot \Delta r \qquad (2.50)$$

For the previous example, at 100 μm distance, (2.50) yields 0.032 K when the input transistors are 20 μm spaced.

Since for a constant drain current, V_{GS} of a MOS transistor varies with temperature with about -2 mV/K [11], the offset voltage of the amplifier varies in time. The variation at double signal frequency has an amplitude of

$$|V_{os}(2\omega_o)| \;=\; 2mV/K.\;\;|\Delta T(2\omega_o)| \tag{2.51}$$

Combining (2.47), (2.49), (2.50) and (2.51) yields:

$$HD_2 \;=\; \frac{|V_{os}(2\omega_o)|}{V_{in}} \;=$$

$$2mV/K.\frac{1}{4\pi\lambda}.\left|\frac{1}{r}+(j+1)\sqrt{\frac{\omega_o}{\delta}}\right|.\frac{\exp(-\sqrt{\omega_o/\delta}.r)}{r}.\Delta r.\frac{|Y_L(\omega_o)|}{|\beta(\omega_o)|}.V_o \tag{2.52}$$

where β is the feedback factor.

As an example, this expression is plotted in Fig. 2.9. versus the distance r, for an amplifier driving a 100 Ohm resistor with 2 Volt output amplitude at 50 Hz, with a feedback factor of 1/3.5 and with Δr equal to 20 μm. For example, when the distance between the output stage and the input transistors equals 100 μm, the second harmonic distortion is —76 dB. This figure illustrates the importance of thermal feedback for low-distortion power amplifier design.

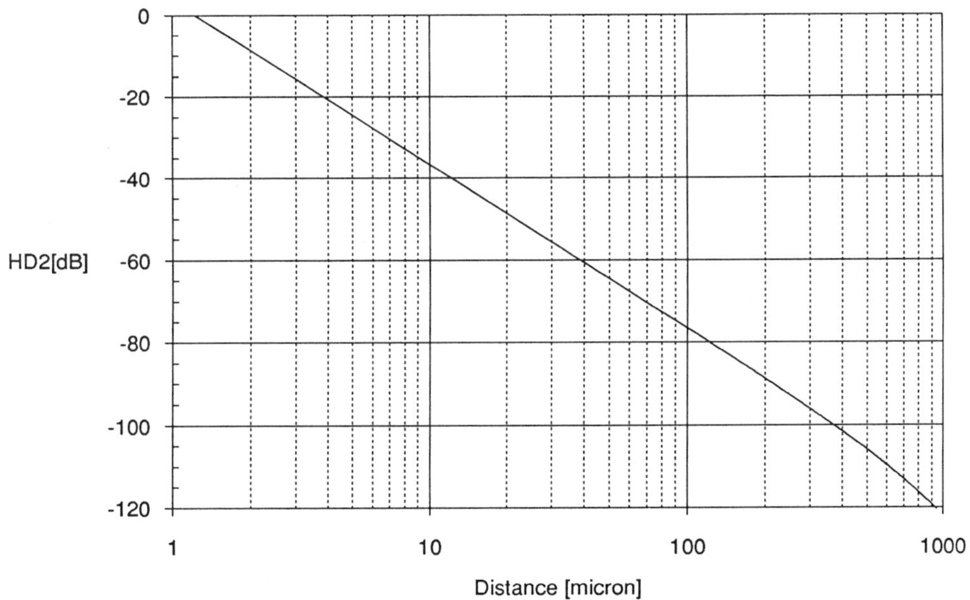

*Fig. 2.9. The second harmonic distortion vs. distance
between the input stage and the output stage*

2.6. A DESIGN EXAMPLE: A CMOS LOW-DISTORTION CLASS AB POWER AMPLIFIER

For dynamic circuits that should exhibit a nearly linear signal transfer characteristic, there are numerous design techniques to minimise the distortion: the open-loop distortion can be reduced, feedback can be applied to futher improve the circuit linearity and, when even this is not sufficient, multiple feedback loops - sometimes also denoted as **local feedback** - can be built in.

Since it is not obvious to catalog all these design techniques and to describe them in a general, abstract way, some distortion minimisation techniques will be treated here by means of a design example: in this section, the design of a low-distortion CMOS Class AB power amplifier, intended for ISDN applications, will be analised.

In Fig. 2.10.a, the principle schematic of a Class AB output stage is depicted. Fig. 2.10.b. shows the idealised transistor currents versus the output current. As opposed to a Class A amplifier, a Class AB amplifier can never be built with linear elements only, since the transistor currents are nonlinear functions of the output current, and thus of the input voltage. So, where the component nonlinearities are undesired second order effects in a Class A amplifier, some of them are essential in a Class AB amplifier. It is important to ensure that these nonlinearities do not degrade the amplifier distortion performance.

A first important issue in the design of such a low-distortion amplifier is the reduction of the open-loop gain nonlinearities and the Cross-over distortion. Therefore, a Class AB output stage with improved distortion performance is described first. Then, a technique is discussed to further reduce this distortion by means of multiple feedback. It will be shown that this distortion reduction is so effective that the remaining harmonic distortion is determined by a second-order effect, similar to the effects described in sections 2.3 to 2.5.

2.6.a. An improved Class AB output stage

The idealised transistor currents as depicted in Fig. 2.10.b. cannot be realised in practice. For most Class AB output stages, the transistor currents are as shown in Fig. 2.11.a. [12-16]. Both output transistors cut off every signal period and on some internal nodes, the impedance becomes high. As a result, it takes a rather large time to switch the polarity of the output current. This deteriorates the cross-over distortion characteristics, especially at high frequencies. Also, when switching the input from its maximum negative to maximum positive value (or vice versa), the slew rate can be poor. A bipolar solution to overcome this problem was proposed by Seevinck *et al* [17] where the current through each output transistor is limited to a minimum value and must never become too small. This is depicted in Fig. 2.11.b.

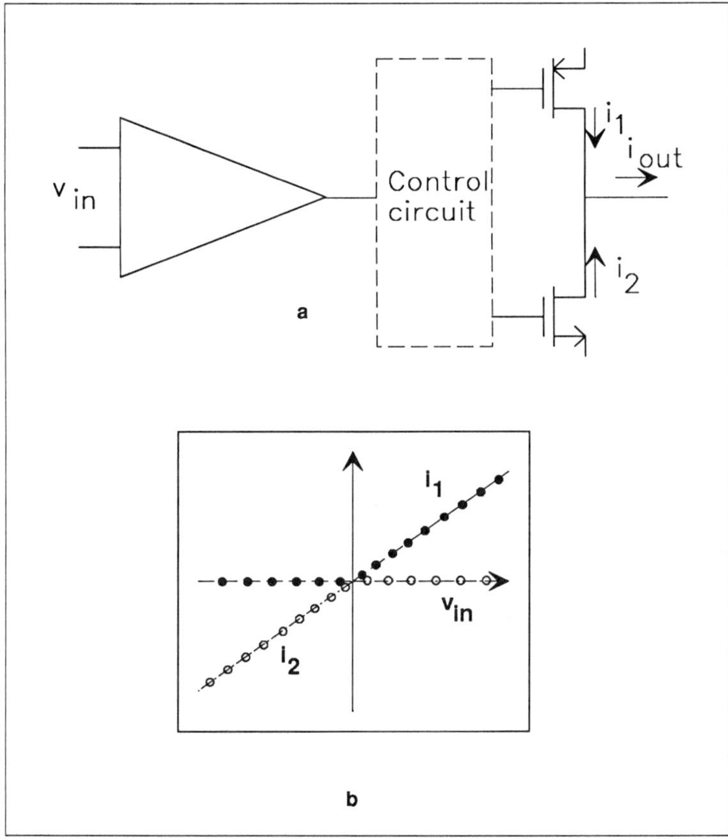

Fig. 2.10.   *a) The principle of a Class AB amplifier*
                 b) Idealised output transistor currents vs. input voltage

In Fig. 2.12.a., the principle schematic of the improved Class AB output stage is depicted. The output stage consists of transistors M_1 and M_2 with input nodes A and B. In the ideal case, the output currents should be functions of the input voltage, as depicted in Fig. 2.12.b.:

1) The smaller one of the currents through M_1 or M_2 has to equal the requested quiescent current and must never become too small.

2) The larger one of the currents through M_1 and M_2 has to be function of the input voltage.

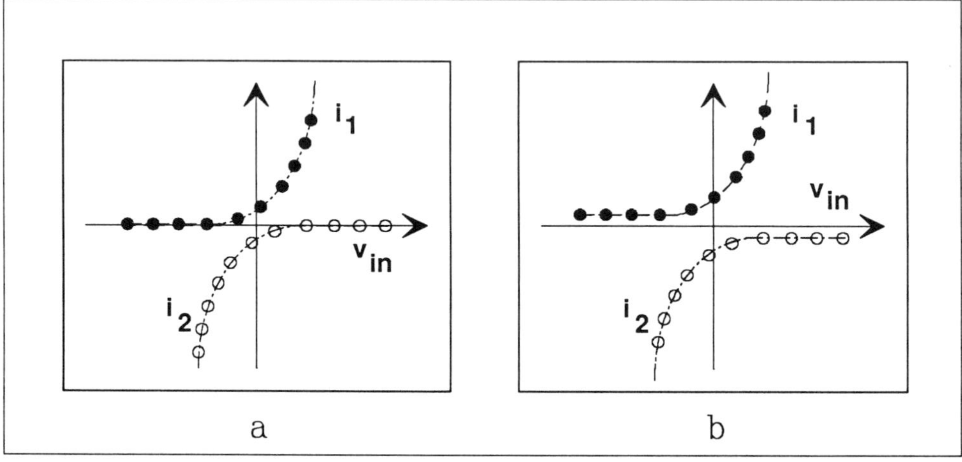

Fig. 2.11. Real output transistor currents vs. input voltage
a) for a classical Class AB output stage
b) for the improved output stage

The basic idea of this output stage is that the common-mode component of the currents through M_1 and M_2 (this determines the quiescent current) can be controlled by changing the differential voltage between nodes A and B, and that the differential mode component of these currents (this is the output current) can be adapted by varying the common mode voltages of A and B.

Both conditions 1) and 2) mentioned above are established by the intermediate amplifier stage which has differential inputs and outputs (see Fig. 2.12.a.). A voltage summator circuit is added which measures the gate voltages of the output transistors (M_1 directly and M_2 via M_3 and M_4) and thus their currents. The result is compared with the reference current through M_{15}.

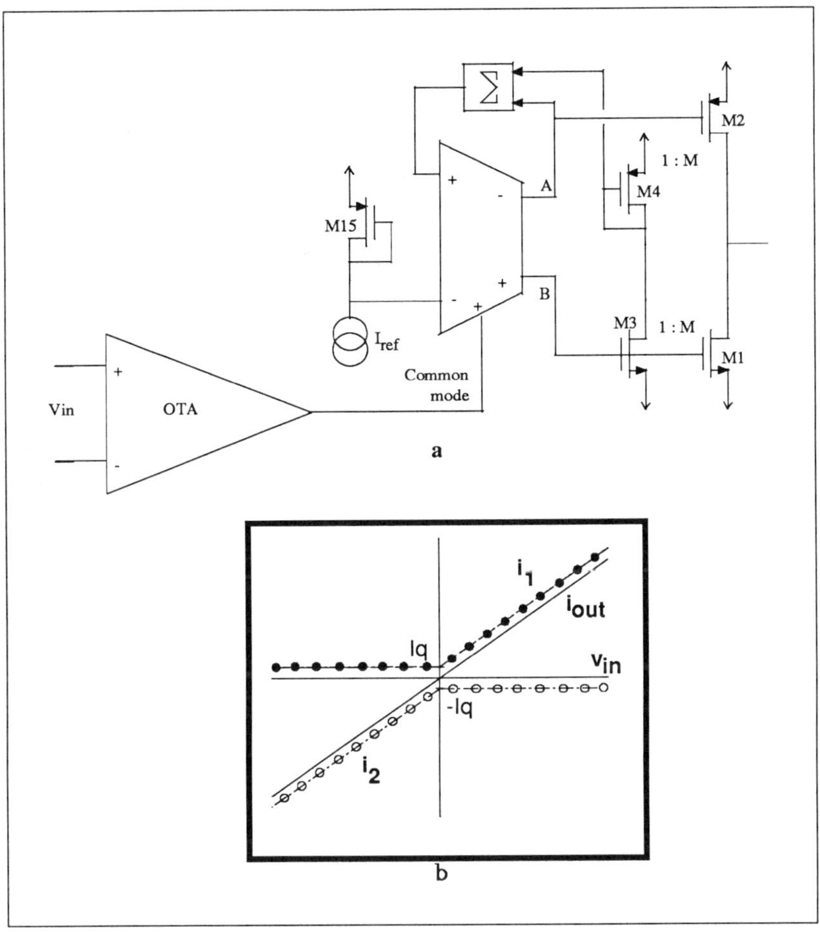

*Fig. 2.12. a) Principle of the described Class AB output
stage b) Idealised output transistor currents*

As mentioned above, a Class AB amplifier always includes some nonlinear circuits. The summator of Fig. 2.12.a. is formed by the circuitry in the dashed zone of Fig. 2.13. This summing circuit is strongly nonlinear because it gives more weight to the output transistor with the smaller current. This can be explained as follows: under quiescent conditions, when M_1 and M_2 carry the same current, the gate voltage of M_{11} and M_{12} are the same. The circuit in the dashed zone of Fig. 2.13. will force these gate voltages equal to the gate voltage of M_{15}. This implies that the quiescent current through M_1 and M_2 is related to the reference current.

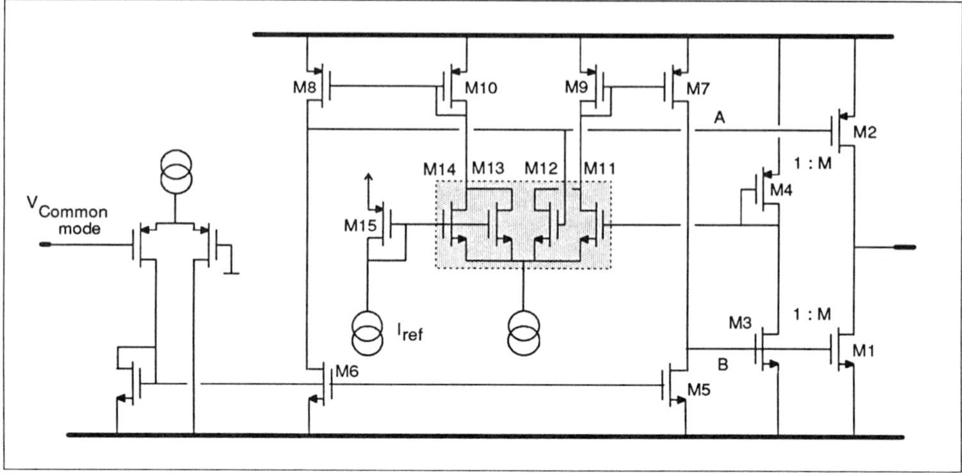

Fig. 2.13. Detail of the Class AB output stage

When for instance a negative voltage is applied to the common-mode input of Fig. 2.12.a, the voltages on nodes A and B will decrease and M_2 will carry a larger current than M_1. Under this non-quiescent condition, the gate voltage of M_4 will become much lower than the gate voltage of M_2. M_{12} will therefore cut off. The circuit in the dashed zone of Fig. 2.13. forces the current through M_{11} equal to the sum of the current through M_{14} and M_{15}. This condition controls the gate voltage of M_4 and thus the current through M_1. In the same way, when M_1 carries a much larger current than M_2, the current through M_2 will be controlled. This illustrates that the circuit in the dashed zone of Fig. 2.13. controls the smaller of the currents through M_1 and M_2. Its nonlinear behaviour is well suited to create the output currents of a Class AB amplifier. Note that M_{11} and M_{12} never cut off at the same time; the impedance on none of the internal nodes becomes too high.

If we assume that no mismatches occur in the transistors M_1 to M_{14} in Fig. 2.13., that transistors M_1 and M_2 satisfy

$$(\mu . C_{ox} . W/L)_1 = (\mu . C_{ox} . W/L)_2 \qquad (2.53)$$

and if a simple square-law transistor model is assumed, the DC transfer characteristic of the output stage can be calculated:

$$\frac{i_{out}}{I_q} = -4 \frac{v_c}{(v_{gs}-V_T)_1} . \{1 + k - \sqrt{k^2 - [v_c/(V_{GS}-V_T)_1]^2}\}$$

$$\text{if } |v_c/(V_{GS}-V_T)_1| < k/\sqrt{2}$$

$$= -4 \; \frac{v_C}{(V_{GS}-V_T)_1} \cdot \left\{ \left| \frac{v_C}{(V_{GS}-V_T)_1} \right| + 1 - k \cdot (\sqrt{2}-1) \right\} \qquad (2.54)$$

$$\text{if } \left| v_C/(V_{GS}-V_T)_1 \right| > k/\sqrt{2}$$

with $k = (V_{GS}-V_T)_{11}/(V_{GS}-V_T)_1$

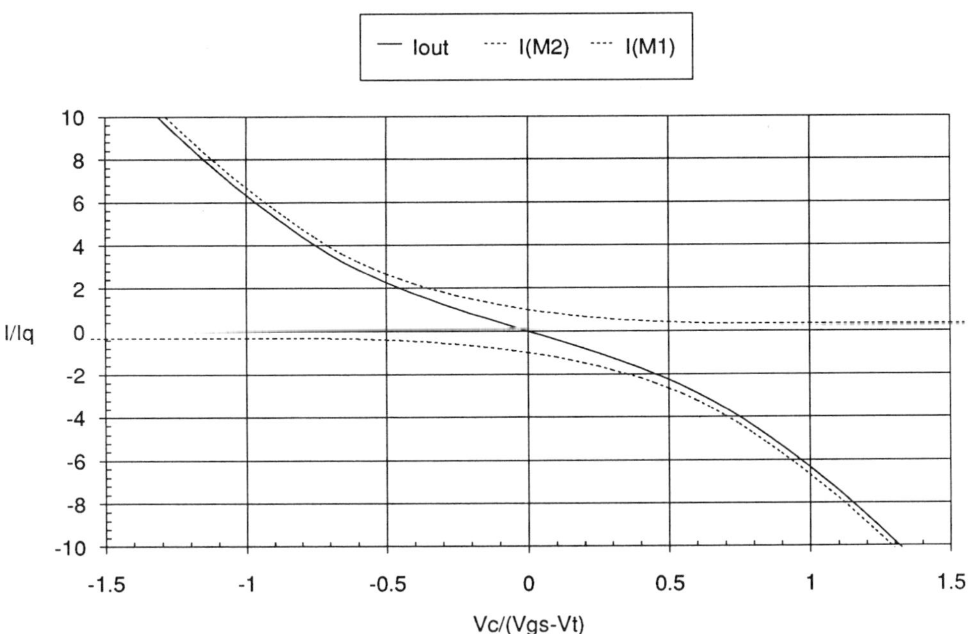

Fig. 2.14. The DC transfer characteristic of the Class AB output stage.

In this expression, v_C is the common mode voltage on node A and B of Fig. 2. 12.a. I_q, $(V_{GS}-V_T)_1$ and $(V_{GS}-V_T)_{11}$ are the quiescent values. In the actual design, parameter k is set to one. This DC transfer characteristic is depicted in Fig. 2.14. along with the currents through the output transistors. When the current through M_1 (M_2) is much larger than the current through M_2 (M_1), the current through M_2 (M_1) will drop to

$$\frac{i}{I_q} = [1 - (\sqrt{2}-1) \cdot k]^2 = 0.34 \text{ for } k = 1 \qquad (2.55)$$

For k smaller than 2.4, the output transistors never cut off.

The harmonic distortion can be calculated from a series expansion of (2.54) and is a function of the ratio $I_{out,peak}/I_q$. Under the assumptions stated above, the DC transfer characteristic is symmetric and there is no second harmonic distortion. According to the technique described in section 2.2.b, the third harmonic can be calculated. For small output currents, the Taylor expansion of expression (2.54) is given by:

$$\frac{i_{out}}{I_q} = -4 \cdot \left\{ \frac{v_c}{(V_{GS}-V_T)_1} + \frac{1}{2} \cdot \left[\frac{v_c}{(V_{GS}-V_T)_1}\right]^3 + \ldots \right\} \qquad (2.56)$$

According to expression (2.9), the third harmonic distortion equals:

$$HD_3 = \frac{1}{128} \cdot \left[\frac{i_{out,\ peak}}{I_q}\right]^2 \qquad (2.57)$$

In Fig. 2.15, the Third harmonic distortion is plotted versus the peak output current. This curve increases with I_{out}^2 at low current, but becomes flat at higher values.

In practice however, the DC transfer function is never perfectly symmetric and second harmonic distortion will occur. Moreover, the output transistors do not satisfy a simple square law model and even come out of saturation for large output currents. Therefore, an accurate calculation of the harmonics is hard to perform. In next section, it will be shown that the distortion due to the output stage will be suppressed by three feedback loops and that other distortion components will dominate. Therefore, an accurate estimation of the output stage distortion is not important.

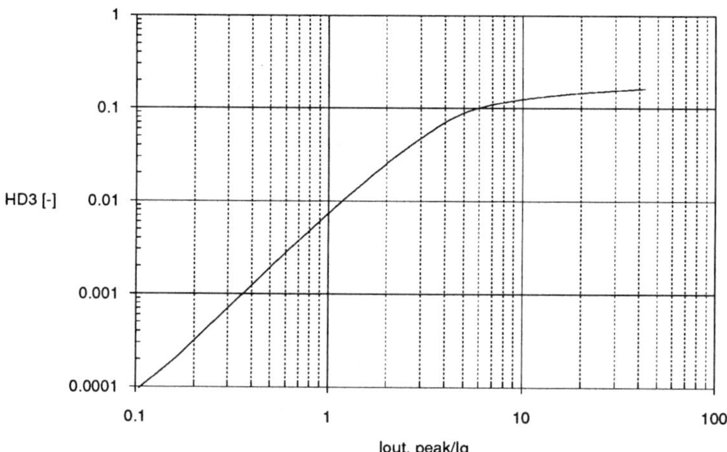

*Fig. 2.15. The calculated Open loop Third harmonic
distortion of the Class AB output stage*

2.6.b. Improving the distortion with three amplifier stages

As demonstrated by expressions (2.32) and (2.33), the harmonic distortion of a system with feedback is suppressed by the loop gain, measured at the considered harmonic frequency. A large loop gain requires a large DC gain and a large GBW.

The amplifier of Fig. 2.13. contsists of two stages: one from the input node to the nodes A and B, and one from these nodes to the output. A small-signal equivalent of this amplifier is shown in Fig. 2.16.a. The DC gain of the output stage is given by:

$$A_{DC} = g_{m2} \cdot R_L \tag{2.58}$$

Due to the heavy resistive load, this DC gain is only in the order of three. When the gain of the first stage is estimated in the order of 100, the loop gain is only in the order of 300. This is not enough to sufficiently suppress the harmonic distortion and to obtain low-distortion performance.

A solution for this problem is depicted in Fig. 2.12.a. This amplifier consists of three stages: one from the input to the node labeled "common mode", one from this node to the nodes A and B and a third stage from these nodes to the output. The small-signal equivalent is shown in Fig. 2.16.b. Two Miller compensation capacitors, C_C and C_D are required to prevent instability. In this section, it is shown that this amplifier type has a higher DC-gain and superior distortion characterisitics compared with the two-stage Miller-compensated amplifier of Fig. 2.16.a.

It is easy to verify that the transfer function of the three-stage amplifier of Fig. 2.16.b. is given by:

$$\frac{V_{out}}{V_{in}} = - \frac{g_{m1} \cdot (g_{m3} \cdot g_{m2} - s \cdot g_{m2} C_D - s^2 \cdot C_C \cdot C_D)}{g_L g_{o1} g_{o2} + s \cdot g_{m2} g_{m3} C_C + s^2 \cdot (g_L + g_{m3}) C_C C_D + s^3 \cdot C_L C_C C_D} \tag{2.59.a}$$

$$\text{provided that} \quad g_{o1}, \ g_{o2} \ll g_{m1}, \ g_{m2} \ll g_{m3} \tag{2.59.b}$$

$$\text{and} \quad C_1, \ C_2 \ll C_C, \ C_D, \ C_L \tag{2.59.c}$$

The numerator of this expression has two zeros, the smaller of which is located in the right s-domain half plane. By making the ratios g_{m2}/C_C and g_{m3}/C_D large enough, these zeros can be located at frequencies well above the GBW. The denominator has three poles located in the left s-domain half plane. By increasing the ratios g_{m2}/C_D and g_{m3}/C_L, the non-dominant poles can be located well above the GBW.

In the same way as for a two-stage amplifier where the stability can be improved by increasing the transconductance of the output stage, the stability condition of a three-stage amplifier requires minimum values for the transconductances of the second and third stage. Compared with the two-stage amplifier of Fig. 2.26.a, the three-stage amplifier has one extra left-half plane zero and one extra pole. The stabilisation of this amplifier does not cause additional difficulties.

The GBW, which is equal to $g_{m1}/2\pi C_C$, is of the same order as the GBW of a two-stage amplifier. The DC-gain however is much larger since there is one more amplifier stage.

The distortion characteristics of a three-stage amplifier are superior compared with those of a two-stage amplifier. The output stage is inside three nested feedback loops (one via C_D, one via C_C and one via the external feedback). Each loop suppresses the distortion by the loop gain, according to expressions (2.33) and (2.34). As a result, the harmonic distortion due to the output stage is suppressed by the product of three loop gains.

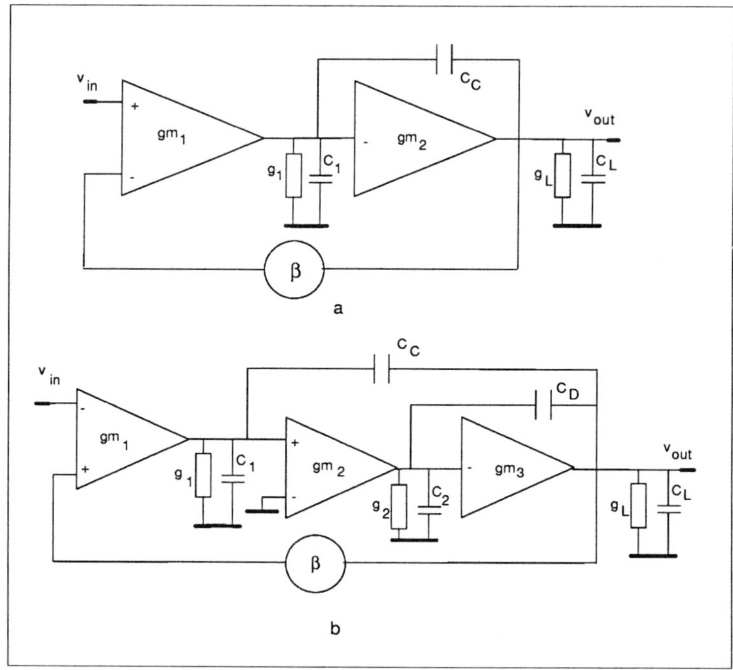

Fig. 2.16: Small-signal equivalent of
a) a Miller-compensated two-stage amplifier and
b) a three-stage amplifier

2.6.c. Realisation and experimental results of an integrated prototype

In Fig. 2.17., the entire schematic of a practical realisation of Fig. 2.12.a. is depicted, intended as a line driver for ISDN applications (See also Fig. 2.13.). M_1 and M_2 form the output stage, M_3 to M_{32} are the intermediate stage and the input stage is formed by transistors M_{36} to M_{44}.

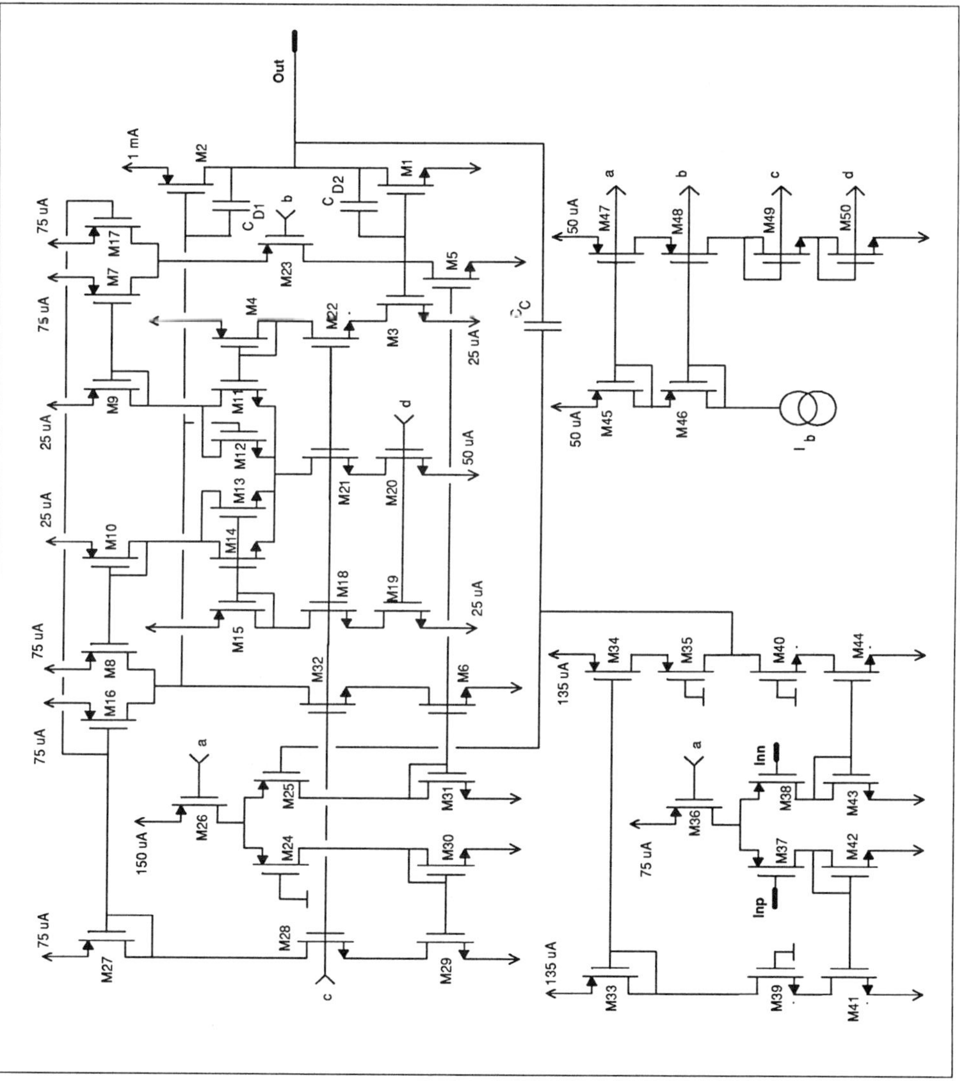

Fig. 2.17. The complete schematic of the Class AB amplifier

Since this design consists of three stages, two compensation capacitors are required as shown in Fig. 2.16.b. The compensation capacitor around the last stage is split up into C_{D1} and C_{D2} in order to ensure stability under non-quiescent conditions. Cascode transistors are used in the input stage, resulting in an over-all DC-gain comparable to a four-stage amplifier. By an appropriate choice of the biasing voltages b and c (See at the bottom right of Fig. 2.17.), the circuit was made as symmetrical as possible. This improves the accuracy of the quiescent current control loop.

The circuit is realised in a standard 2 μm double metal double poly CMOS process. Fig. 2.18. shows a chip microphotograph. The silicon area is 0.28 mm^2, bonding pads not included. During layout, special attention has been paid to the thermal behaviour of the amplifier. The output transistors are placed horizontally in Fig. 2.18. and are covered with large metal strips. This causes the isotherms to be horizontal. Transistors which have to match are placed on the same isotherm. The input transistors form a common-centoid layout in order to further suppress the thermal feedback.

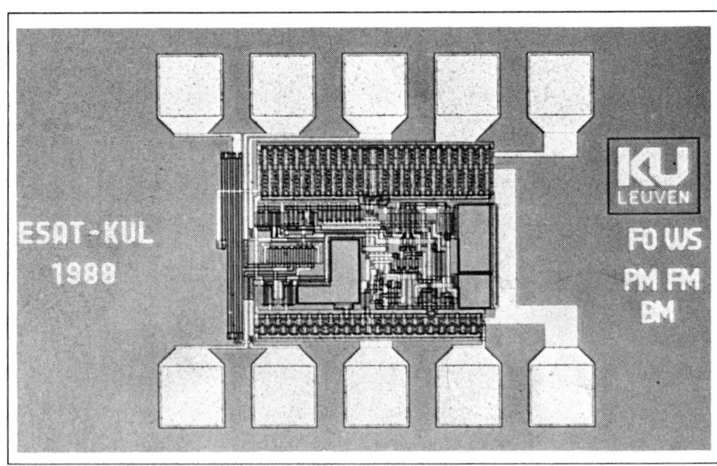

Fig. 2.18. A microphotograph of the Class AB amplifier

In Fig. 2.19., the measured open loop gain and phase are depicted. Note that this amplifier was intended to be used with a feedback factor of 1/3.5. Therefore, the gain-bandwidth of the loop is only 5 MHz. The phase margin has to be evaluated at this 5 MHz. Since the DC gain is so large, it is hard to measure with a network analyser. This explains the inaccuracies in Fig. 2.19. at low frequencies.

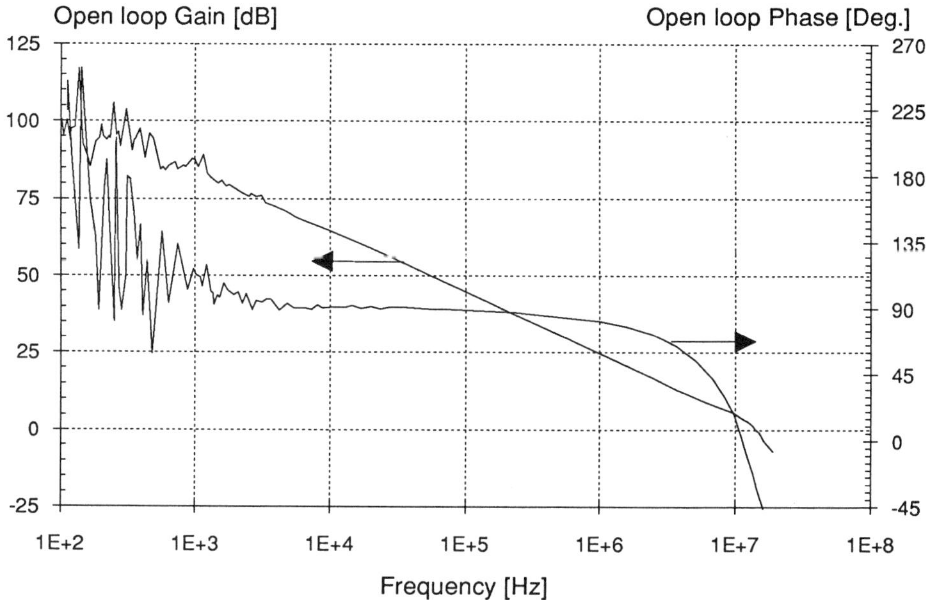

Fig. 2.19. The measured amplitude and phase
characteristics

The measured amplifier characteristics are summarized in Table 2.1. Note the excellent PSRR characterisitcs for the negative power supply: Up to a frequency of 100 kHz, a PSRR over 60 dB is obtained [6].

Supply Voltage		+/- 2.5 V
Quiescent Power Consumption		10.3 mW
Load		81 Ω//15 pF
Output Swing		+/- 2.2 V
Feedback factor		1/3.5
GBW of the loop		4.9 MHz
Phase Margin of the loop		58°
Gain Margin of the loop		7 dB
DC gain		120 dB *
PSRR$^-$	DC	70 dB
	100 kHz	66 dB
	1 MHz	28 dB
PSRR$^+$	DC	37 dB
	100 kHz	39 dB
	1 MHz	28 dB
S/(THD+N)**	1 kHz	82 dB
	10 kHz	80 dB
	50 kHz	55 dB

* simulated

** output amplitude: 1.75 V

all measurements with nominal load and
   supply voltage

Table 1: Measured Class AB amplifier Characteristics

2.6.d. Comparison of the distortion measurements with calculations

The black curve in Fig. 2.20. shows the total harmonic distortion of the amplifier when used as a non-inverting buffer with feedback factor 1/3.5. Up to approximately 30 kHz, the distortion is about -80 dB, rapidly increasing for higher frequencies. The objective of this section is to compare this distortion figure with calculations and to find the major distortion contribution.

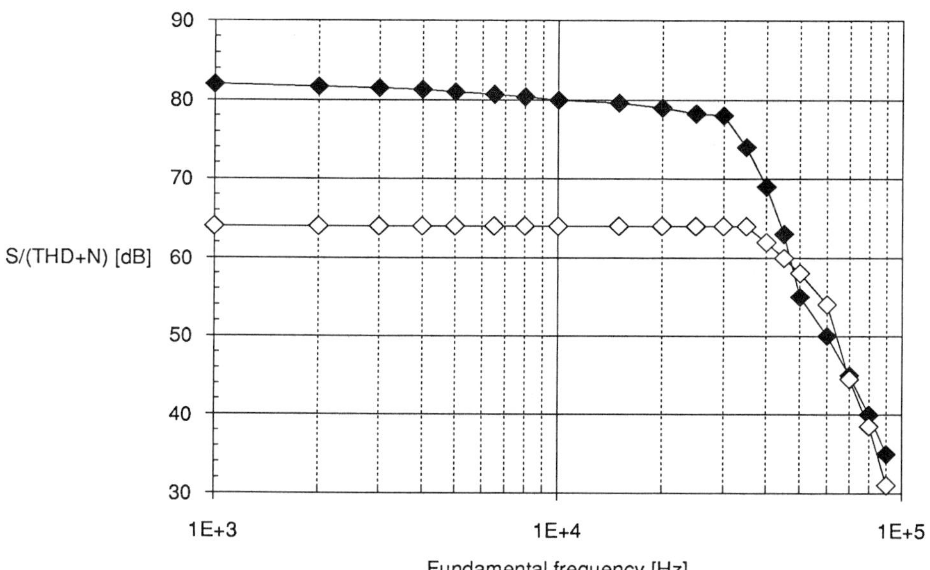

Fig. 2.20: The measured Harmonic distortion vs. Frequency
black: for the amplifier of Fig. 2.17.
white: for the same amplifier with a resistor of 4.7 Ω is series with V_{DD}

The output stage is situated within three nested feedback loops. As mentioned before, its distortion is suppressed by the product of three loop gains, as expressed by (2.32) and (2.33):

$$HD_{2f}(\omega_o) = HD_{2o}/ \left[\frac{\beta \cdot 2\pi GBW}{2\omega_o} \cdot \frac{2\pi \cdot p_2}{2\omega_o} \cdot (gm_1+gm_2) R_L \right] \qquad (2.60)$$

$$HD_{3f}(\omega_o) = HD_{3o}/ \left[\frac{\beta \cdot 2\pi GBW}{3\omega_o} \cdot \frac{2\pi \cdot p_2}{3\omega_o} \cdot (gm_1+gm_2) R_L \right] \qquad (2.61)$$

With HD_{2o} and HD_{3o} in the order of 30%, ω_o equal to, for instance, $2\pi.10$ kHz and p_2 equal to 10 MHz, (2.60) and (2.61) yield —111 dB resp. —104 dB. A comparison with the measured value of -80 dB shows that the output stage is not the major distortion source.

According to (2.17) the distortion due to the first stage equals

$$HD_3 = \frac{1}{64}.\frac{1}{3}.\{\frac{V_{out}}{(V_{GS}-V_T)_{38}}\}^2.\{\frac{\omega_o}{2\pi GBW}\}^2.\frac{3\omega_o}{2\pi GBW\beta} \qquad (2.62)$$

$$= -177 \text{ dB}$$

when $(V_{GS}-V_T)_{38}$ equals 0.2 Volt.

In practice however, the input stage is not symmetric and second harmonic distortion will occur. According to (2.39) and (2.41),

$$HD_2 = \frac{1}{4}.\frac{\beta.V_{out}}{(V_{GS}-V_T)_{38}}.\varepsilon.\frac{1}{[1+T(\omega_o)]^2} = -136 \text{ dB} \qquad (2.63)$$

where ε is determined by mismatches (ε is estimated about 5%). So, this nonlinearity is also not the source of the major distortion contribution. With a similar analysis, it can be calculated that the distortion of the intermediate stage is also far below the measured value of -80 dB.

From the results of previous alineae, it can be concluded that basic distortion calculation techniques fail to predict the measured harmonic distortion characteristics. Even when the nonlinearities of the compensation capacitors or the feedback resistors are considered, no sufficient explanation of the measured distortion can be found.

An explanation for the measured harmonic distortion can be found from the second-order effects described before in sections 2.3. to 2.5: the PSRR for the positive supply voltage is rather low, due to parasitic feedthrough via M_{26} (see Fig. 2.17). In section 2.4, the harmonic distortion due a low PSRR is calculated for a Class A amplifier. For a Class AB amplifier however, the situation is much worse. As shown in Fig. 2.21, each supply voltage delivers output current during approximately one half of the signal period. Therefore, each of the supply currents contains multiple harmonics of the signal frequency. Even a linear power-supply gain feeds these harmonics back to the amplifier input stage and cause distortion:

$$HD_{2k} = \frac{\pi}{2}.\frac{1}{(2k-1).(2k+1)}.\frac{Y_L(\omega_o).Z_S(2k\omega_o)}{\beta(\omega_o)}.\frac{1}{PSRR(2k\omega_o)} \qquad (2.64)$$

$$HD_{2k+1} = 0$$

A supply impedance of 0.3 Ohm is sufficient to generate a second harmonic distortion of -80 dB.

In order to confirm the importance of the power supply gain, the harmonic distortion of the Class AB amplifier is measured with an extra resistor of 4.7 Ohm in series with the positive supply voltage. According to expression (2.64), a harmonic distortion of -57 dB could be expected. The measured results are shown in Fig. 2.20. (white curve). Although there is still a difference of 7 dB between the measured distortion values and the calculations, the two curves of Fig. 2.20 clearly illustrate the importance of the PSRR of a low-distortion amplifier.

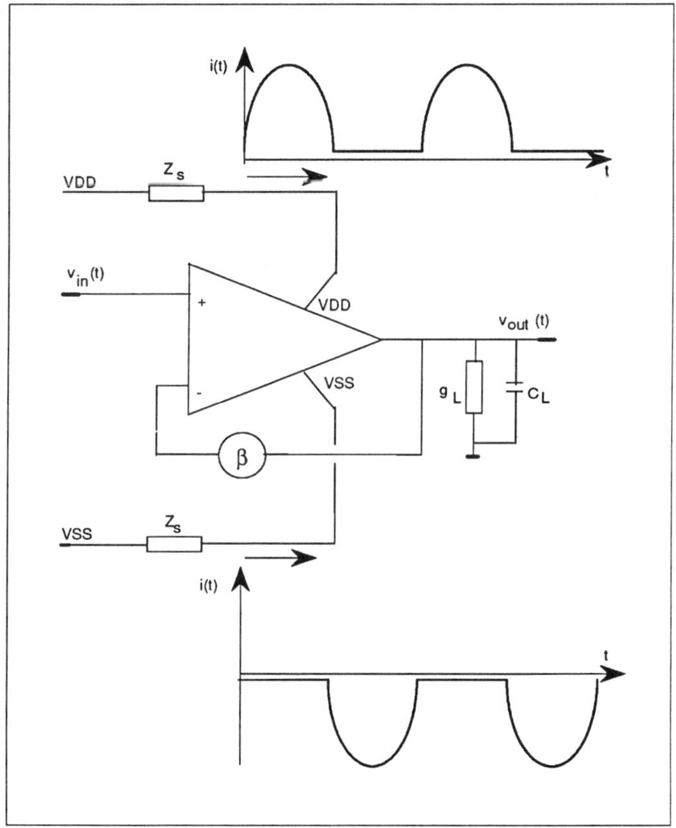

Fig. 2.21: The supply currents of a Class AB amplifier vs. time

2.7. SUMMARY

This chapter deals with several aspects of low-distortion amplifier design. The minimisation of the distortion due to the nonlinear open-loop differential gain is treated by means of a design example. This example incorporates techniques to improve the open-loop gain linearity and it exploits both global and local feedback.

Further, it is shown both analytically and with examples that in practice, most of the distortion is caused by other nonlinearities which are normally treated as second-order effects: for an amplifier used as a non-inverting buffer, the nonlinear common-mode gain causes additional distortion which is not suppressed by feedback. For a power amplifier driving a heavy load, the power supply gain and the thermal feedback have a similar effect.

For a Class AB amplifier, nonlinearities are essential for a proper behaviour. Therefore, the danger for undesired distortion is much more acute. For instance, when the power supply impedance is not zero, these nonlinearities cause nonlinear power supply variations which influence the circuit when the power supply gain is considerable.

This chapter reveals that a low-distortion amplifier not only requires a fairly linear open loop gain and a sufficient DC gain and gainbandwidth, but also requests a large CMRR and PSRR. A careful layout is mandatory because of thermal feedback.

2.8. REFERENCES

[1]   K. LAKER, W. SANSEN: "Design of Analog Circuits and Systems" - to be published by Mc Graw-Hill, 1990

[2]   M. SCHETZEN: "The Volterra and Wiener Theories of Nonlinear Systems" - John Wiley and Sons, New-York, 1980

[3]   S. NARAYANAN: "Application of Volterra Series to intermodulation distortion of Transistor feedback Amplifiers" - *IEEE Trans. on Circuit Theory*, vol, CT-17, No. 4, Nov. 1970 pp. 518-527

[4]   J. J. BUSSGANG, L. EHRMAN, J. W. GRAHAM: "Analysis of Nonlinear Systems with multiple Inputs" - *Proc. of the IEEE,* vol. 62, No. 3, Aug. 1974, pp. 1088-1119

[5]   S. L. GARVERICK, C. G. SODINI: "Large-Signal Linearity of Scaled MOS Transistors" - *IEEE J. Solid-State Circuits* vol. SC-22, No. 2, April 1987, pp. 282-286

[6]   F. OP 'T Eynde *et al.*: "A CMOS Large-Swing low-Distortion Three-stage Class AB Power Amplifier" - *IEEE J. Solid-State Circuits*, vol. SC-25 no. 1, Febr. 1990 pp. 266-273

[7]   F. Op 'T EYNDE, P. WAMBACQ, W. SANSEN: "On the Relationship between the CMRR or PSRR and the Second Harmonic Distortion of Differential Input Amplifiers" - *IEEE J. Solid-State Circuits* vol. SC-24 no.6, Dec. 1989 pp.1740-1744

[8]   Y. P. TSIVIDIS, D. L. FRASER: "Harmonic Distortion in Single-Channel MOS integrated Circuits" - *IEEE J. Solid-State Circuits* vol. SC-16, No. 6, Dec. 1981 pp. 694-702

[9]   M. J. THOMA, W. T. BAUMANN, C. R. WESTGATE: "A Method to Predict Harmonic Distortion in Small-Geometry MOS Analog Integrated Circuits" - *IEEE J. Solid-State Circuits* vol. SC-22, No. 1, Febr. 1987 pp.106-109

[10]   J. E. SOLOMON: "The Monolithic Opamp: a Tutorial Study" - *IEEE J. Solid-State Circuits* vol. SC-9, Nov. 1974 pp.314-323

[11]   P. R. GRAY *et al*: "Threshold voltage temperature drift in Ion-Implanted MOS Transistors" - *IEEE J. Solid-State Circuits* Vol. SC-17, No. 2, April 1982, pp. 291-298

[12]   R. CASTELLO, P. R. GRAY: "A High-Performance Micropower Switched-Capacitor Filter" - *IEEE J. Solid-State Circuits* vol. SC-20, no. 6, pp. 1122-1132, Dec. 1985

[13]   S. L. WONG, C. A. T. SALAMA: "An Efficient Buffer for Driving Large Capacitive Loads" - *IEEE J. Solid-State Circuits* vol. SC-21, No. 3, June 1986, pp. 464-469

[14]   K. E. BREHMER, J. B. WIESER: "Large Swing CMOS Power Amplifier" - *IEEE J. Solid-State Circuits* vol. SC-18, No. 6, June 1983, pp. 624-628

[15]   J. A. FISHER: "A High-Performance CMOS Power Amplifier" - *IEEE J. Solid-State Circuits* vol. SC-20, No. 6, June 1985, pp. 1200-1205

[16]   L. CALLEWAERT, W. SANSEN: "Class AB CMOS Amplifiers with High Efficiency" - *IEEE J. Solid-State Circuits* vol SC-25, no. 3, June 1990

[17]   A. SEEVINCK, W. DE JAGER, P. BUITENDIJK: "A Low-Distortion Output Stage with Improved Stability for Monolithic Power Amplifiers" - *IEEE J. Solid-State Circuits* vol. SC-23, No. 3, June 1988, pp. 794-801

APPENDIX 2.A: SOME REMARKS ABOUT THE VOLTERRA SERIES

When a nonlinear time-invariant system is driven with an input signal x(t), the output signal y(t) is a nonlinear function of the system input signal. Let us denote the fourier transform of x(t) as X(jω). According to the Volterra Series theory [2], the relationship between x(t) and y(t) can be described as follows:

$$y(t) = H_0 + \frac{1}{2\pi}\cdot\int H_1(\omega_1).X(j\omega_1).\exp(j\omega_1 t).d\omega_1 \quad +$$

$$\frac{1}{(2\pi)^2}\cdot\iint H_2(\omega_1,\omega_2).X(j\omega_1).X(j\omega_2).\exp(j\omega_1 t).\exp(j\omega_2 t) \; d\omega_1.d\omega_2 \quad +$$

$$\frac{1}{(2\pi)^3}\cdot\iiint H_3(\omega_1,\omega_2,\omega_3).X(j\omega_1)X(j\omega_2)X(j\omega_3)\exp(j\omega_1 t)\\ .\exp(j\omega_2 t)\exp(j\omega_3 t).d\omega_1.d\omega_2.d\omega_3 \quad +$$

$$\ldots \tag{2.65}$$

where all integrals range from $-\infty$ to $+\infty$. The coefficients $H_k(\omega_1,\omega_2,...,\omega_k)$ are the **Volterra kernels** [2]. These Volterra kernels describe the nonlinear behaviour of the system: H_0 is the output signal when no input signal is applied. $H_1(\omega)$ describes the small-signal linearised behaviour of the system; it is the linearised ac-gain, commonly used in ac analyses. The terms with $H_2(\omega_1,\omega_2)$, $H_3(\omega_1,\omega_2,\omega_3)$, $H_4(\omega_1,\omega_2,\omega_3,\omega_4)$, etc. describe the circuit nonlinearities. Expression (2.65) is valid, provided that this series reaches convergency.

Note also the following property of the volterra kernels:

$$H_k(\omega_1,\omega_2,\ldots,\omega_k) \text{ is the complex conjugate of } H_k(-\omega_1,-\omega_2,\ldots,-\omega_k) \tag{2.66}$$

Consider now the situation where the nonlinear system is driven by a sinusoidal signal:

$$x(t) = V_{in}.\cos(\omega_0 t), \text{ and hence,}$$

$$X(j\omega) = V_{in}.\pi.[\delta(\omega-\omega_0) + \delta(\omega+\omega_0)] \tag{2.67}$$

where $\delta(\cdot)$ represents the Dirac impulse function.

According to expression (2.65), the output signal is then given by:

$$y(t) = H_0 +$$

$$V_{in}/2.\{ H_1(\omega_o).\exp(j\omega_o t)+H_1(-\omega_o).\exp(-j\omega_o t) \} \qquad +$$

$$V_{in}^2/4.\{ \begin{array}{l} H_2(\omega_o,\omega_o)\exp(2j\omega_o t)+H_2(-\omega_o,\omega_o)+ \\ H_2(\omega_o,-\omega_o)+H_2(-\omega_o,-\omega_o)\exp(-2j\omega_o t)+ \} \end{array} \qquad +$$

$$V_{in}^3/8.\{ \begin{array}{l} H_3(\omega_o,\omega_o,\omega_o)\exp(3j\omega_o t)+H_3(-\omega_o,\omega_o,\omega_o)\exp(j\omega_o t)+ \\ H_3(\omega_o,-\omega_o,\omega_o)\exp(j\omega_o t)+H_3(\omega_o,\omega_o,-\omega_o)\exp(j\omega_o t)+ \\ H_3(-\omega_o,-\omega_o,\omega_o)\exp(-j\omega_o t)+H_3(-\omega_o,\omega_o,-\omega_o)\exp(-j\omega_o t)+ \\ H_3(\omega_o,-\omega_o,-\omega_o)\exp(-j\omega_o t)+ \\ H_3(-\omega_o,-\omega_o,-\omega_o)\exp(-3j\omega_o t) \} \end{array} \qquad +$$

$$V_{in}^4/16.\{ \begin{array}{l} H_4(\omega_o,\omega_o,\omega_o,\omega_o)\exp(4j\omega_o t)+ \\ H_4(\omega_o,\omega_o,\omega_o,-\omega_o)\exp(2j\omega_o t)+ \\ H_4(\omega_o,\omega_o,-\omega_o,\omega_o)\exp(2j\omega_o t)+ \\ H_4(\omega_o,-\omega_o,\omega_o,\omega_o)\exp(2j\omega_o t)+ \\ H_4(-\omega_o,\omega_o,\omega_o,\omega_o)\exp(2j\omega_o t)+ \\ H_4(\omega_o,\omega_o,-\omega_o,-\omega_o)+H_4(\omega_o,-\omega_o,\omega_o,-\omega_o)+ \\ H_4(-\omega_o,\omega_o,\omega_o,-\omega_o)+H_4(\omega_o,-\omega_o,-\omega_o,\omega_o)+ \\ H_4(-\omega_o,\omega_o,-\omega_o,\omega_o)+H_4(-\omega_o,-\omega_o,\omega_o,\omega_o) \\ H_4(-\omega_o,-\omega_o,-\omega_o,\omega_o)\exp(-2j\omega_o t)+ \\ H_4(-\omega_o,-\omega_o,\omega_o,-\omega_o)\exp(-2j\omega_o t)+ \\ H_4(-\omega_o,\omega_o,-\omega_o,-\omega_o)\exp(-2j\omega_o t)+ \\ H_4(\omega_o,-\omega_o,-\omega_o,-\omega_o)\exp(-2j\omega_o t)+ \\ H_4(-\omega_o,-\omega_o,-\omega_o,-\omega_o)\exp(-4j\omega_o t) \} \end{array}$$

$$+ \ldots \qquad\qquad\qquad\qquad\qquad\qquad\qquad\qquad\qquad\qquad (2.68)$$

The output y(t) consists of a DC component, of a fundamental component with frequency ω_o and of components at multiples of the input signal frequency.

As can be seen, the DC component of y(t) consists of a contribution from H_0, of a contribution from $H_2(\omega_o,-\omega_o)$, of a contribution proportional with $H_2(-\omega_o,\omega_o)$ and of contributions due to H_4. In general, all the even-order Volterra kerners contribute to the DC componemt of y(t). However, under the assumption of **low distortion**, these contributions decrease rapidly with increasing kernel order. Only the contribution due to H_0 is significant.

Similarly, the second harmonic component in y(t) consists of contributions due to H_2, due to H_4 and due to all other even-order Volterra kernels. Under the assumption of low distortion, only the contribution due to H_2 is significant.

In the same way, under the assumption of low distortion, only the contributions due to H_1 have to be considered to calculate the fundamental signal. In general, the k-th harmonic can be calculated by considering only the terms due to H_k.

Therefore, under low distortion conditions, expression (2.68) reduces to:

$$y(t) = H_0 +$$

$$V_{in}/2 . \{ H_1(\omega_o) . \exp(j\omega_o t) + H_1(-\omega_o) . \exp(-j\omega_o t) \} \qquad +$$

$$V_{in}^2/4 . \{ H_2(\omega_o, \omega_o) \exp(2j\omega_o t) + H_2(-\omega_o, -\omega_o) \exp(-2j\omega_o t) \} \qquad +$$

$$V_{in}^3/8 . \{ H_3(\omega_o, \omega_o, \omega_o) \exp(3j\omega_o t) + $$
$$H_3(-\omega_o, -\omega_o, -\omega_o) \exp(-3j\omega_o t) \} \qquad +$$

$$V_{in}^4/16 . \{ H_4(\omega_o, \omega_o, \omega_o, \omega_o) \exp(4j\omega_o t) + $$
$$H_4(-\omega_o, -\omega_o, -\omega_o, -\omega_o) \exp(-4j\omega_o t) \}$$

$$+ \ldots \qquad (2.69)$$

When taking the characteristic (2.66) into account, this huge expression (2.69) can be rewritten as

$$y(t) = \sum_k \frac{V_{in}^k}{2^k} . \{ H_k(\omega_o, \omega_o, \ldots, \omega_o) . \exp(kj\omega_o t) + $$
$$H_k^*(\omega_o, \omega_o, \ldots, \omega_o) . \exp(kj\omega_o t) \} \qquad (2.70)$$

where the asterix stands for the compex conjugate. This can again be reformulated to:

$$y(t) = \sum_k V_k . \cos(k\omega_o t + \varphi_k) \qquad (2.1.b)$$

where $\qquad V_k = \dfrac{V_{in}^k}{2^{k-1}} . | H_k(\omega_o, \omega_o, \ldots, \omega_o) |$

$$\text{and } \varphi_k = \arg(H_k(\omega_o, \omega_o, \ldots, \omega_o)) \qquad (2.71)$$

Comparing this expression with expression (2.5.a) shows that the coefficients A_k in (2.5.a) are nothing else than the k-th order Volterra kernels $H_k(\omega_o, \omega_o, \ldots, \omega_o)$. Expressions (2.1.b) and (2.71) allow to calculate the harmonic distortion, once the Volterra kernels are known.

To calculate these Volterra kernels in an easy way, consider the (mathematical) situation where the input signal is given by:

$$x(t) = \exp(j\omega_o t) \quad \text{and hence,} \quad X(j\omega) = 2\pi . \delta(\omega - \omega_o) \qquad (2.72)$$

According to expression (2.65), the output signal y(t) is then given by:

$$y(t) = H_0 + \int H_1(\omega_1) \cdot \delta(\omega_1 - \omega_o) \cdot \exp(j\omega_1 t) \cdot d\omega_1 \quad +$$

$$\iint H_2(\omega_1, \omega_2) \cdot \delta(\omega_1 - \omega_o) \cdot \delta(\omega_2 - \omega_o) \cdot \exp(j\omega_1 t) \cdot \exp(j\omega_2 t) \, d\omega_1 \cdot d\omega_2 \quad +$$

$$\iiint H_3(\omega_1, \omega_2, \omega_3) \cdot \delta(\omega_1 - \omega_o) \cdot \delta(\omega_2 - \omega_o) \cdot \delta(\omega_3 - \omega_o) \cdot \exp(j\omega_1 t) \cdot \exp(j\omega_2 t) \exp(j\omega_3 t) \cdot d\omega_1 \cdot d\omega_2 \cdot d\omega_3 \quad +$$

$$\dots \tag{2.73}$$

Or, after calculating the integrals:

$$y(t) = H_0 + H_1(\omega_o) \cdot \exp(j\omega_o t) + H_2(\omega_o, \omega_o) \cdot \exp(2j\omega_o t) +$$

$$H_3(\omega_o, \omega_o, \omega_o) \cdot \exp(3j\omega_o t) + \dots \tag{2.74}$$

and, since the coefficients A_k in expressions (2.4) and (2.5) are nothing else than the k-th order Volterra kernels:

$$y(t) = \sum_k A_k \cdot \exp(jk\omega_o t) \tag{2.4.b}$$

This proves the correctness of the distortion calculation technique for dynamic circuits.

3 OVERSAMPLED A-TO-D AND D-TO-A CONVERTERS

3.1. INTRODUCTION

Digital signal processors convert digital input signals into digital output signals. But since most real-life signals have an analog nature, analog-to-digital and digital-to-analog converters are necessary to interface between a DSP and the outside world. In practice, these converters are the bottle necks of the over-all system: the analog-to-digital conversion rate limits the signal bandwidth while the word length determines the signal-to-noise ratio. Both the analog-to-digital and digital-to-analog converters introduce analog component noise that degrades the signal quality. In high-performance data acquisition systems, the data converters are therefore the critical building blocks.

The resolution of classical data converters based on binary-weighted arrays [1-23] is limited by component matching properties. Low-noise performance requires expensive trimming procedures which are not tolerated in a mass production environment. Oversampled data converters form an alternative where a high resolution, independent of component matching properties, is obtained at the cost of a speed reduction and some additional digital postprocessing.

The first application of oversampling techniques can be found about thirty years ago in the Delta modulators for digital data transmission applications [24]. In order to decrease the sensitivity for offset and slope overload, the Delta modulator architecture was further improved towards a Sigma-Delta modulator. Later [25] [26], it was found that the Sigma-Delta principle could also be applied for A-to-D and D-to-A converters.

Although most of the system aspects of a Sigma-Delta modulator are well-known by now, there is no general text book about this topic. The state of the art is divided over more than one hundred papers. In the two following chapters, the design requirements for Sigma-Delta A-to-D and D-to-A converters are studied. A thoroughful understanding of the basic principles of a Sigma-delta modulator are essential to understand these chapters. Therefore, this chapter provides a general introduction to the principle of Sigma-Delta modulation and a summary of the available literature.

Because of major differences between the characteristics of a Delta modulator and a Sigma-Delta modulator, the Sigma-Delta modulator is considered in this text without referring to the original Delta modulator. The principle and its application for data converters are demonstrated. Relationships are presented between the accuracy and the speed penalty.

The oversampling technique is compared with alternative techniques. It is shown that the Sigma-Delta principle allows to realise cost-effective data converters with a high accuracy (20 bit or even better) at a moderate sampling rate (up to a few 100 kHz).

3.2. ANALOG SIGNALS VERSUS DIGITAL SIGNALS

Analog signals are continuous-time, continuous-amplitude signals while digital signals have a discrete-time, discrete-amplitude nature. Converting an analog signal into a digital equivalent therefore requires two steps, as depicted in Fig. 3.1.a.: first, the analog signal is **sampled** in order to obtain a discrete-time, continuous-amplitude signal.

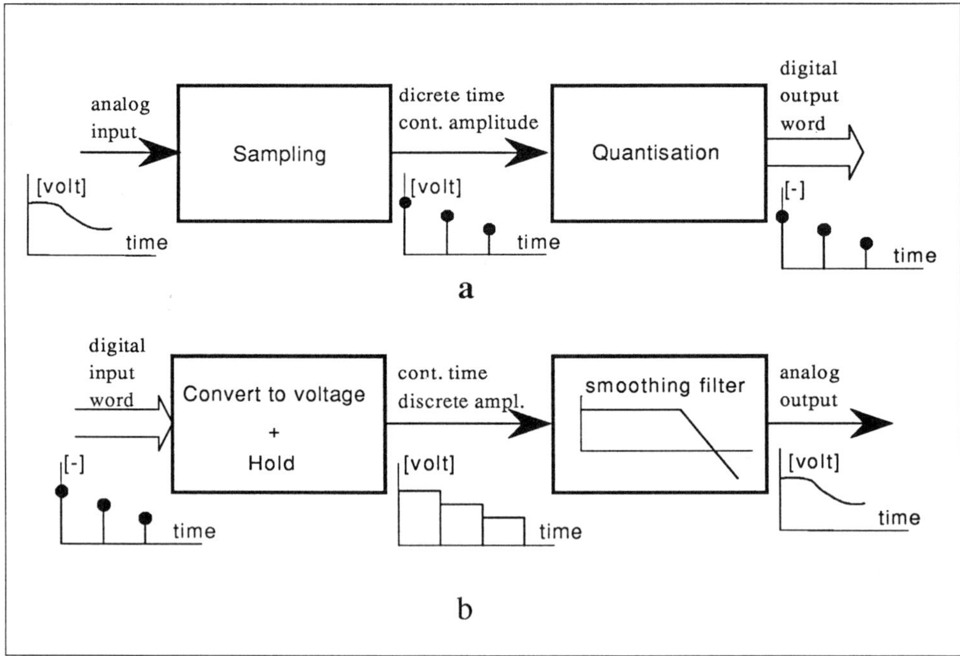

Fig. 3.1: The principle schematic of a) an ADC and b) a DAC

In a second step, this intermediate signal is **quantised** to obtain a discrete amplitude. The conversion of a digital signal into an analog one basically consists of a **hold** operation followed by a **smoothing** lowpass filter as demonstrated in Fig. 3.1.b.

In this section, the four basic operations involved - sampling, quantisation, hold and smoothing - are discussed more in detail. Their consequences for the signal degradation are investigated.

3.2.a. The sampling

In Fig. 3.2.a, an analog signal and its frequency spectrum are depicted. The sampled signal and its spectrum are presented in Fig. 3.2.b. The sampling produces side bands of the signal band around every multiple of the sampling frequency f_s [27]. When the original signal contains contributions at frequencies higher than half of the clock frequency, the baseband is disturbed by the nearest side bands. This is denoted as **aliasing** [27]. But when the sampling rate is larger than two times the highest analog signal frequency, the side bands are well separated, as depicted in Fig. 3.2.c. The baseband can be retrieved with a lowpass filter.

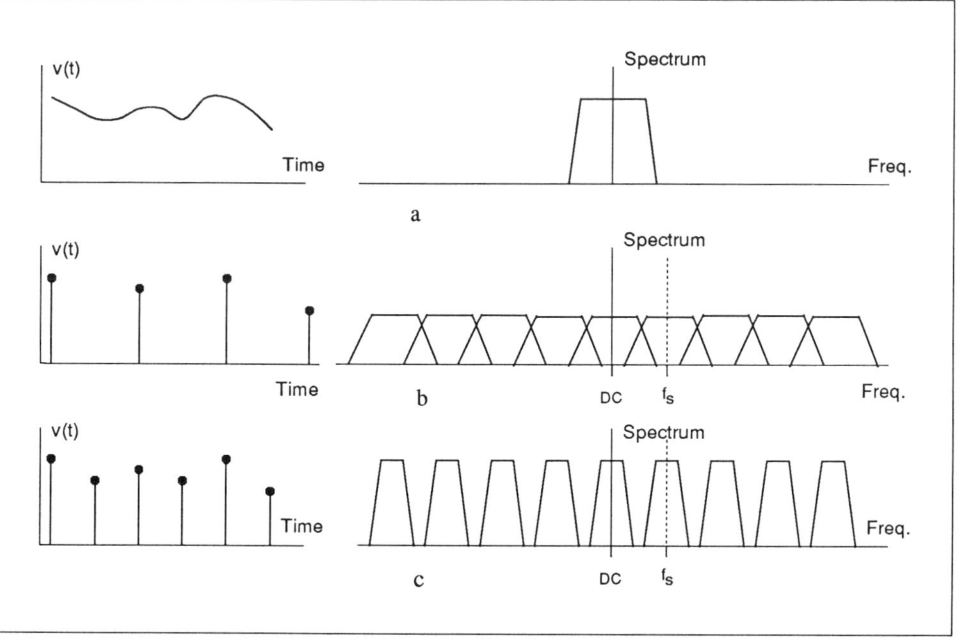

Fig. 3.2: a) an analog signal b) undersampled and c) properly sampled

Hence, sampling causes no signal degradation provided that the analog signal is band limited and that the sampling rate is at least two times the highest analog signal frequency.   This is the well known **Nyquist criterion** and the minimum sampling frequency is called the **Nyquist rate** [27].

There is no signal degradation when the sampling rate is too high: the resulting sampled signal will contain more samples than necessary, but the resulting signal information will be the same. For instance, the signals of Fig. 3.3. all contain the same information, although the sampling rate of Fig. 3.3.c. is larger than that of Fig. 3.3.b. This implies that there is redundancy in the signal of Fig. 3.3.c.

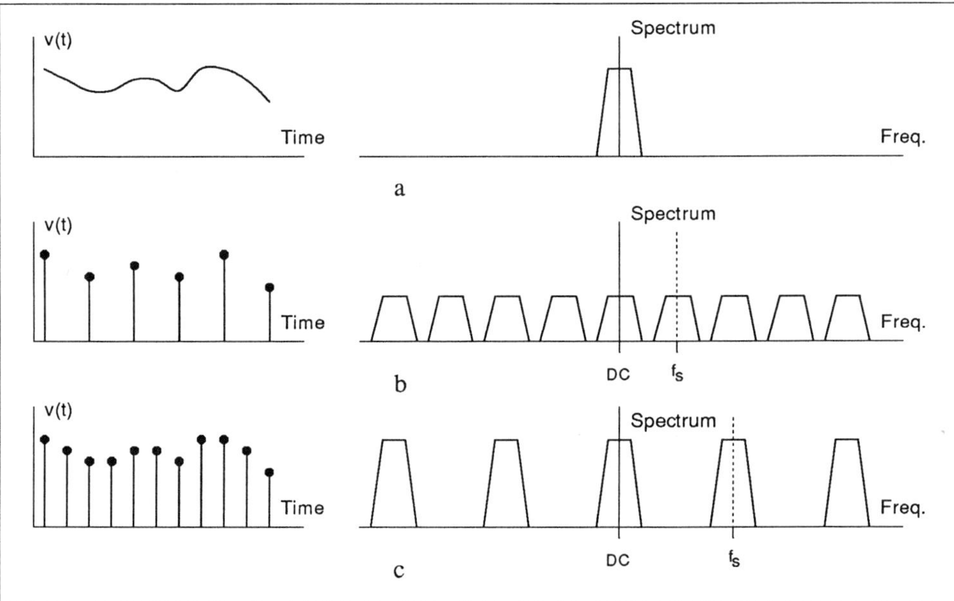

Fig. 3.3: Spectra of a) an analog signal and b,c) sampled equivalents

The **Oversampling Ratio** (OR) of a sampled signal is defined as

$$OR = \frac{\text{actual sampling rate}}{\text{Nyquist rate}} \qquad (3.1)$$

and should be equal to at least one. E.g. for the signal of Fig. 3.3.b, the OR is about one. For the signal of Fig. 3.3.c, it equals two.

Since the signals of Fig. 3.3.b. and Fig. 3.3.c. contain the same information, it is possible to convert one into the other. The process of increasing the clock frequency (converting Fig. 3.3.b. into Fig. 3.3.c.) is called **interpolation** while decreasing the sampling rate (converting Fig. 3.3.c. into Fig. 3.3.b.) is denoted as **decimation** [28].
The principle of interpolation is depicted in Fig. 3.4. When for instance the sampling frequency of a signal has to be increased with a factor of two, from a lower frequency f_d to a higher frequency f_s, the signal with the frequency spectrum of Fig. 3.4.a. has to be converted into a signal with a spectrum as depicted in fig. 4.3.c. In a first step, dummy samples equal to zero are inserted to increase the sampling rate. This is called **upsampling**. For interpolation with more than a factor of two, more dummy samples have to be inserted.

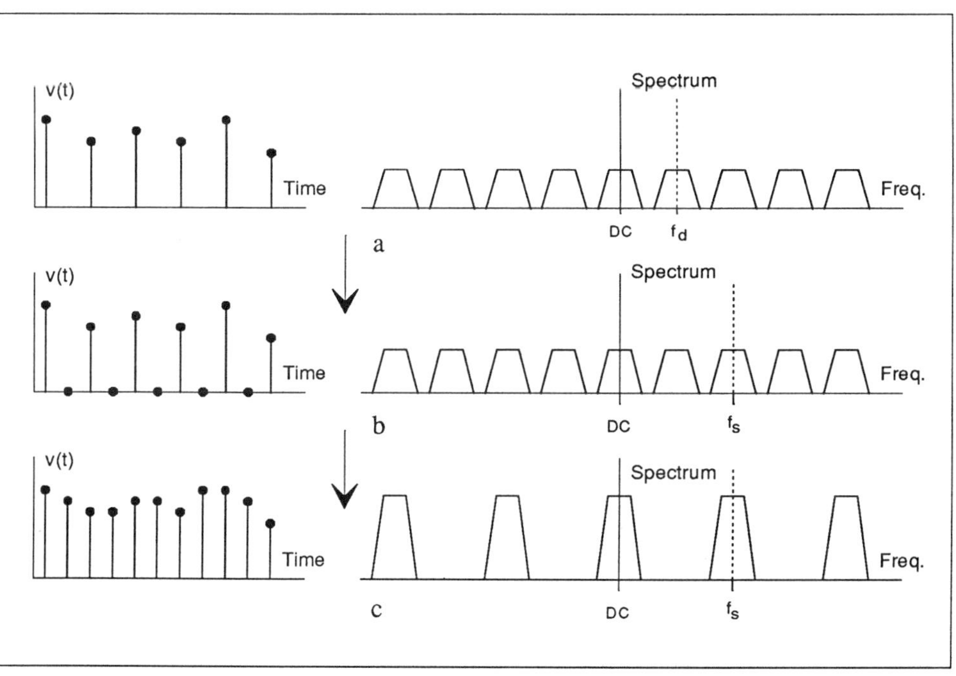

Fig. 3.4: The interpolation mechanism

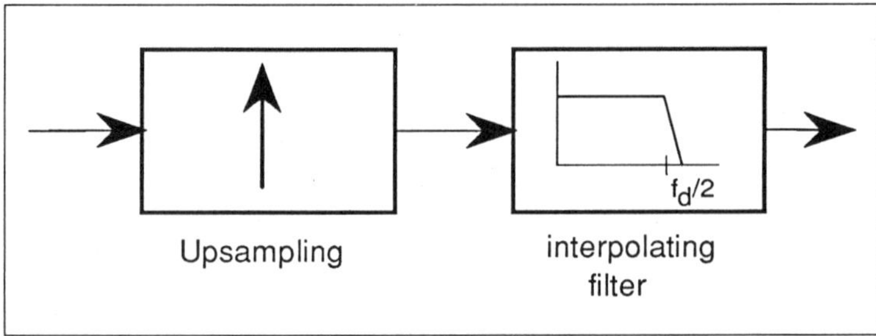

Fig. 3.5: The symbolic notation of an interpolation

The resulting signal (Fig. 3.4.b.) has the requested sampling rate but the frequency spectrum contains undesired side bands at frequencies below half of f_s. By lowpassing the result, the requested signal without undesired side bands (Fig. 3.4.c.) is obtained. The symbolic notation of an interpolation is depicted in Fig. 3.5. The box with the arrow pointing up performs the upsampling. The lowpass filter with cut-off frequency $f_d/2$ used in the interpolation is an **interpolating filter**.

When the Oversampling Ratio of a sampled signal is larger than one, the clock frequency can be decreased with a decimation. The decimation process is depicted in Fig. 3.6. When for instance the sampling rate has to be decreased with a factor of two from f_s to f_d, one of every two samples is deleted, resulting in the requested signal with half of the original sampling rate. This process is denoted as **downsampling**. Decimation with more than a factor of two can be performed by deleting more samples. However, when the original signal contains spurious signals (for instance noise) at frequencies higher than half of f_d, they will be aliased into the signal band at the downsampling as depicted in Fig. 3.6.b. Therefore, a lowpass filter is normally included before the downsampler to remove the signal components at frequencies above half of f_d. This filter is a **decimating filter**. The block schematic of a decimator is depicted in Fig. 3.7. The box with the arrow pointing down performs the downsampling.

Interpolation or decimation can change the sampling rate with an integer factor only. When the sampling rate has to be increased with for instance a factor of 1.5, an interpolation with a factor of three has to be performed first, followed by a decimation with a factor of two.

It can be concluded that the sampling operation causes no signal degradation provided that the sampling rate is at least equal to the Nyquist rate. In practice, it has to be a little larger because the lowpass filters mentioned above are not infinitely steep.

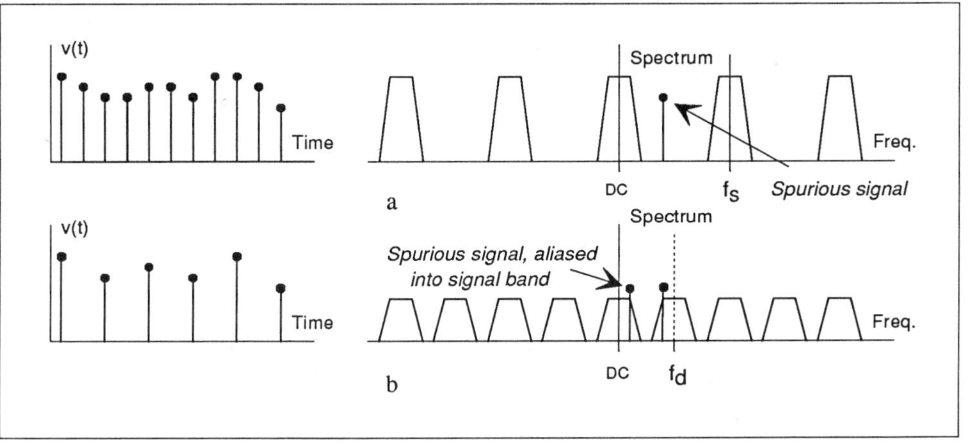

Fig. 3.6: The decimation principle
a) Signal before downsampling
b) Signal after downsampling

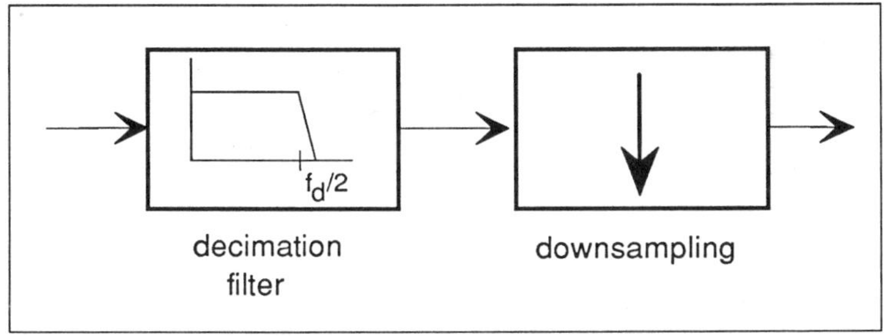

Fig. 3.7: The symbolic notation of a decimation

The clock signal determining the sampling is not perfect but has some jitter. As a result, the sampling instants are not exactly equidistant in time but have some uncertainty as demonstrated in Fig. 3.8. When the input signal is not constant in time, sampling at an incorrect time causes an error signal. For the special case of a sinusoidal input with frequency f and amplitude A, the error can be calculated:

$$\Delta v = A.\cos(2\pi f.t).\omega.\Delta t \tag{3.2}$$

where Δt is the timing error. This error appears as noise signal, called **aperture noise**. The RMS value equals [29]:

$$\sigma^2(\Delta v) \;=\; \frac{1}{2}.A^2.(2\pi f)^2.\sigma^2(\Delta t) \tag{3.3}$$

$$\text{and } SNR = -20.\log[2\pi f.\sigma(\Delta t)] \text{ dB} \tag{3.4}$$

where σ stands for the standard deviation. For instance with a signal frequency of 100 kHz, the SNR equals 104 dB when the timing inaccuracy is 10 psec.

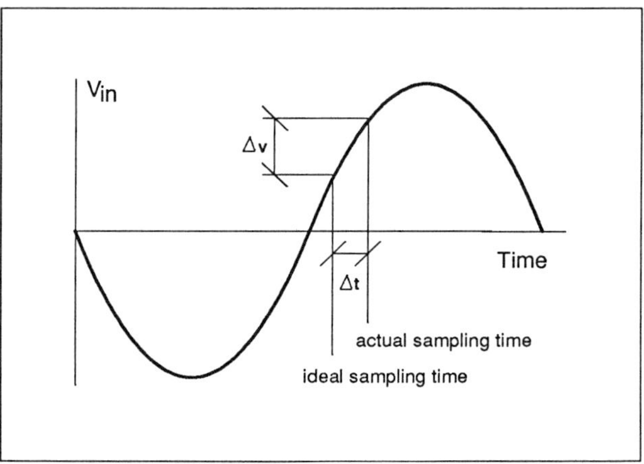

Fig. 3.8: SNR degradation by clock jitter

3.2.b. The quantisation

The sampling transforms the analog input into a discrete-time, continuous-amplitude signal. In the quantiser, this signal is compared with a reference voltage and the ratio (a number with no dimension) is truncated to a finite word length.

The transfer characteristic of a quantiser is depicted in Fig. 3.9.a. As can be seen, the truncation is a nonlinear operation causing an error signal as shown in Fig. 3.9.b. Although a quantisation is a deterministic operation (when we know what goes in, we know what comes out), the error is often treated as statistical, white noise. Strictly mathematical, this error is an intermodulation product of the input signal and the clock, produced by the hard nonlinearities of the transfer chacteristic depicted in Fig. 3.9.a. and it is correlated with the input signal. However, when the input signal varies sufficiently in time, the quantisation noise varies almost randomly (see Fig. 3.9.b.), producing a quasi-continuous power spectrum.

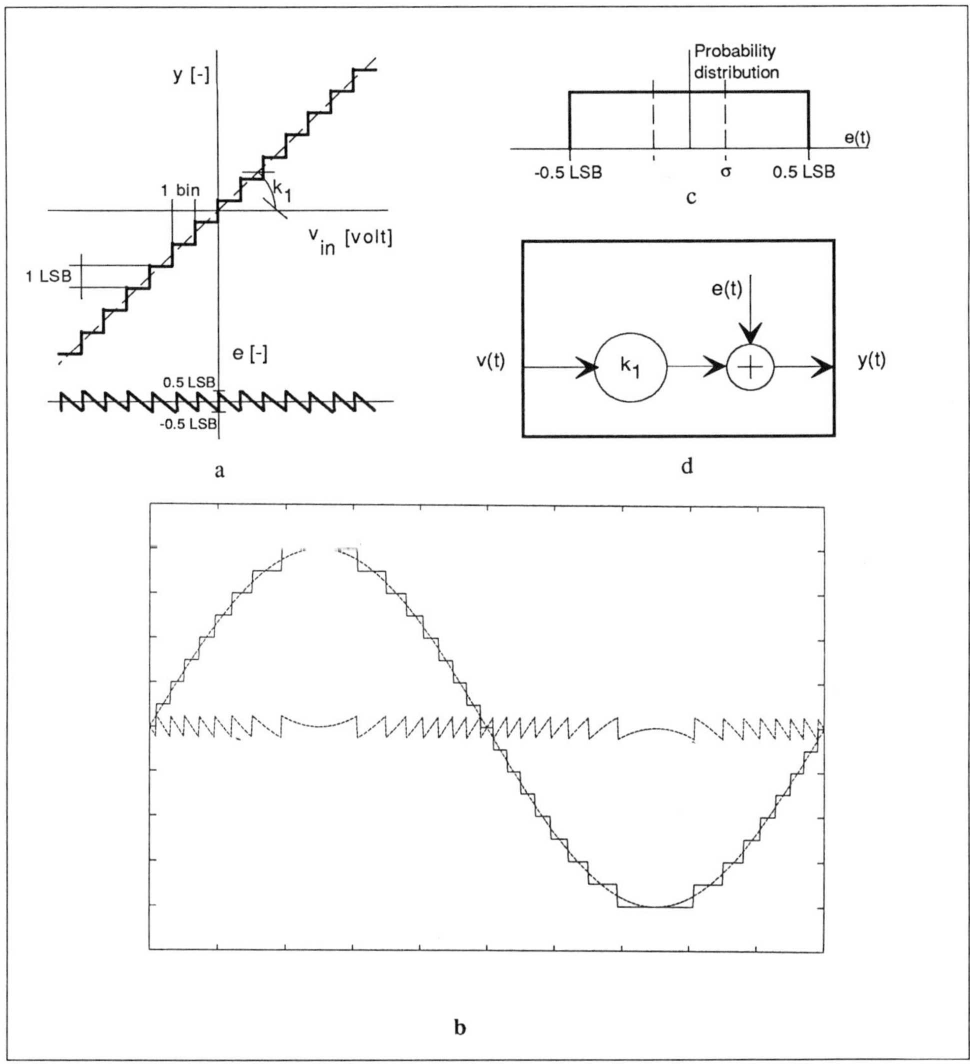

Fig. 3.9:   *a) The Quantisator Transfer characteristic*
         b) Quantisation of a sinus
         c) Probability density function of the error
         d) The equivalent schematic

The range of input signals resulting in the same digital output is a **bin** (see Fig. 3.9.a.). For an ideal quantiser, all the bins have equal size. The resolution of the digital output equals one **Least Significant Bit** (LSB). A quantiser can be modelled by the equivalent schematic of Fig. 3.9.d: the output is proportional to the input, but an error $e(t)$, resulting in quantisation noise is added. The error signal is between -0.5 LSB and

+0.5 LSB, with a flat amplitude distribution as depicted in Fig. 3.9.c. Its RMS value equals [30]:

$$\sigma^2(e) = 1/3 * (0.5 \text{ LSB})^2 \qquad\qquad (3.5)$$

If the word length after the quantisation equals B bits, the maximum amplitude of the input signal is 2^{B-1} bins and for a sinusoidal input signal, the SNR is given by [30]:

$$SNR = 10.\log[\frac{(2^{B-1} \text{ LSB})^2/2}{1/3 * (0.5 \text{ LSB})^2}] = 1.78 + 6.B \text{ dB} \qquad (3.6)$$

Quantisation causes a signal degradation resulting in **quantisation noise**. E.g. for an 8 bit ADC, the SNR equals 50 dB while for an 16 bit ADC, the SNR is 98 dB. High-performance, low-noise ADCs therefore require a large output word length.

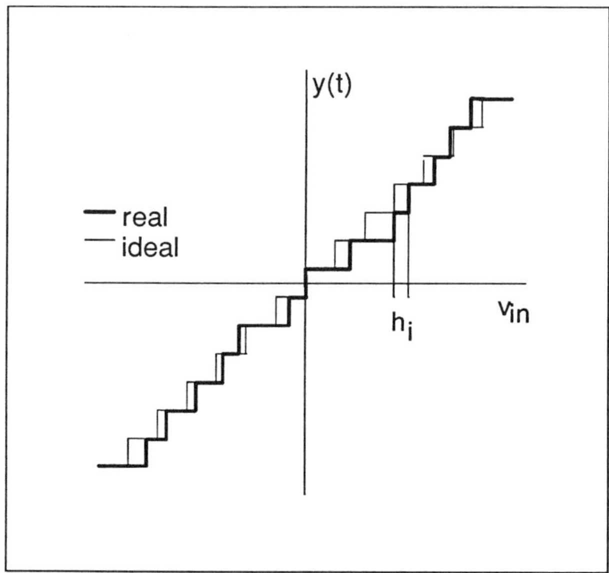

Fig. 3.10: Transfer characteristic of a real quantiser

Up till now, it has been assumed that the bins in Fig. 3.9.a. all have equal size. For a real ADC however, the transfer characteristic is as depicted in Fig. 3.10. The deviation h_i of the i-th stair from its ideal position is the **Integral Nonlinearity of the i-th stair**. The standard deviation of these errors $\sigma^2(h_i)$ is the **Integral Nonlinearity**

(INL) of the ADC. For the non-ideal quantiser, the quantisation error becomes approximately (*):

$$\sigma^2(e) = 1/3*(0.5 \ LSB)^2 + \sigma^2(h_i) \tag{3.7}$$

For example, when the INL is a half bin, the additional signal degradation is 6 dB.

From (3.7) it can be concluded that the quantisation noise is a function of both the word length and the INL. Increasing the word length in order to improve the SNR is useful only when the position of the stairs in the transfer characteristic can be accurately controlled. When the INL is larger than a half bin, the least bit is not significant.

3.2.c. The Hold and the Smoothing operations

A DAC converts the digital input signal into an analog voltage, as shown in Fig. 3.1.b. When the conversion from the digital signal to the voltage is nonlinear, additional noise is generated by the DAC:

$$N = \sigma^2(h_i) \tag{3.8}$$

where h_i is the integral nonlinearity, similar to the INL of an ADC.

In order to convert the discrete-time digital signal into a continuous value, a DAC keeps the analog output constant during a clock period (see fig. 3.1.b.). This is a linear **hold** operation with transfer function

$$H(f) = \frac{1}{f_s} . \exp(-j\pi.f/f_s) . \frac{\sin(\pi.f/f_s)}{\pi.f/f_s} \tag{3.9}$$

where f is the signal frequency and f_s is the sampling rate. The signal spectra before and after the DAC are depicted in Fig. 3.11. The high-frequency side bands can be removed by a lowpass filter, called a **smoothing** filter.

As demonstrated in Fig. 3.11, the amplitude characteristic of the hold is not flat. For a maximum signal frequency of $f_s/2$, the hold operation attenuates the signal with a factor of 0.63. A flat frequency spectrum can only be obtained by a preprocessing of the digital input signal.

(*)This noise includes the signal degradation due to harmonic distortion. It is sometimes denoted as "N+THD".

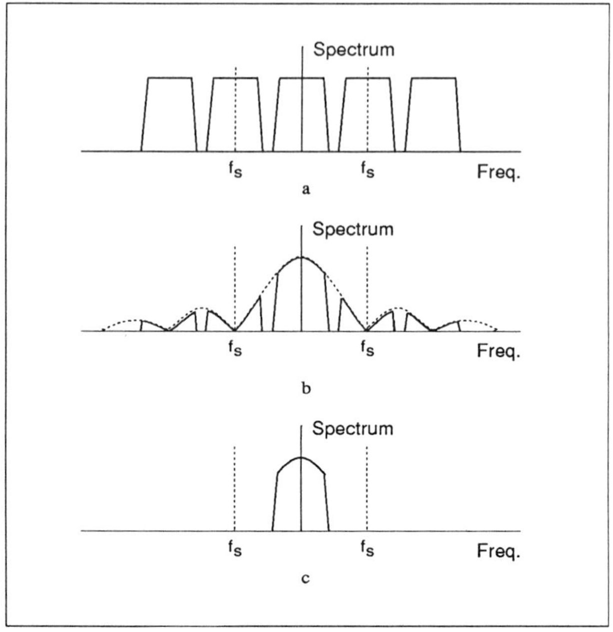

Fig. 3.11: Frequency Spectrum degradation by the hold
a) Signal before the hold
b) Signal after the hold
c) Signal after the smoothing filter

3.2.d. Conclusions

The signal bandwidth of a data converter is determined by the sampling rate: a sampling rate of at least twice the highest input signal frequency is required. The SNR is determined by the clock jitter, the word length and the INL. The INL should be below a half bin. This requirement is the most difficult one: with modern short-channel CMOS processes, sampling rates of a 100 Msample/sec. can be achieved [3]. Also, implementing an ADC with the complexity of for instance 32 bit output word length is no problem, except for the fact that the positions of the stairs have to be accurate within a half bin.

Classical ADCs based on component matching (for instance a flash [1-5] or a pipelined ADC [14-16]) are therefore limited to about 10 to 12 bit. Improved noise performance with these converter types can only be achieved with trimming (which is expensive) or with self-calibration techniques [54]. The same conclusion holds for classical DACs relying on component matching (for instance DACs with binary-weighted arrays).

In the next section, a technique is discussed to overcome the limitations of matched components and to obtain resolutions in the order of 14 to 20 bit without any matching requirements.

3.3. THE PRINCIPLE OF OVERSAMPLED DATA CONVERTERS

As stated in the previous section, the INL requirements for an ADC or DAC are hard to meet for a word length higher than 10 bit. On the other hand, modern short-channel CMOS processes offer a speed performance which is often far beyond the requirements. Moreover, since shorter channel lengths will be available in the future, speed will further improve. The accuracy and component matching however are expected to become worse.

Hence, it would be interesting to trade off speed for accuracy and obtain some accuracy advantage at the cost of a speed limitation. As an example, let us consider the following problem: suppose a signal with 12.5 kHz bandwidth has to be processed with 11 bit accuracy. Assume further that a 10 bit, 100 ksample/sec ADC is available, although an 11 bit, 25 ks/sec ADC is required. Although the available ADC cannot be applied directly, it might be possible to perform some operation to its output, resulting in a lower output rate (lower time resolution) but a higher word length (higher amplitude resolution).

3.3.a. An Oversampled ADC

One possible solution is depicted in Fig. 3.12. The analog input signal is sampled at a rate f_s (100 ks/s) which is considerably larger than the Nyquist rate f_d (25 ks/s in our example). When the quantisation noise power is assumed to be equally distributed over the frequency spectrum as depicted in Fig. 3.12.b, its power spectral density is given by:

$$E^2(f) = \frac{1}{f_s/2}.\frac{1}{3}.[0.5 \text{ LSB}]^2 \qquad (3.10)$$

The higher the sampling rate, the lower the noise power is per unit of frequency and the less noise power is present in the signal band. An eleventh bit, equal to zero is added to the output and the result is lowpassed with a digital lowpass filter with eleven bit accuracy and a filter frequency f_{LPF} equal to $f_d/2$ (12.5 KHz). As a result, three quarters of the noise power is filtered out and the resulting quantisation noise is equivalent to that of an 11 bit ADC. Since the signal band is below 12.5 kHz, the signal is not disturbed by this operation. Now, the sampling frequency can be downsampled with a factor of four to obtain an output rate f_d of 25 ks/sec. The resulting digital output words have a length of 11 bit which are all significant. An accuracy improvement is obtained at the cost of an intermediate high frequency and some digital postprocessing. The lowpass filter and the downsampler form a decimator.

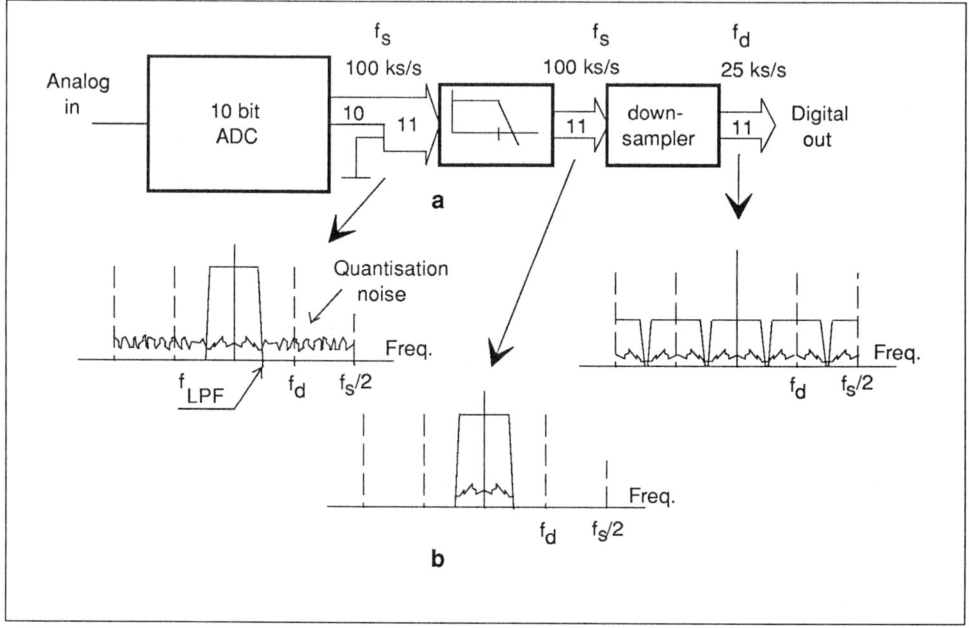

Fig. 3.12: An Oversampled ADC

This technique has however two problems:
1) The trade-off between speed and accuracy is not very advantageous: a speed reduction of a factor of four offers only one bit of accuracy improvement.
2) It is assumed that the quantisation noise of the 10 bit ADC is equally distributed between DC and $f_S/2$. In practice, this condition is not met. For instance when a DC input signal is applied as depicted in Fig. 3.13, the quantisation noise is equal for all the samples (see Fig. 3.13.c.) and its spectrum contains only a DC component. Then the lowpass filter will filter no noise out at all. The output words are 11 bit long but only 10 bit are significant.

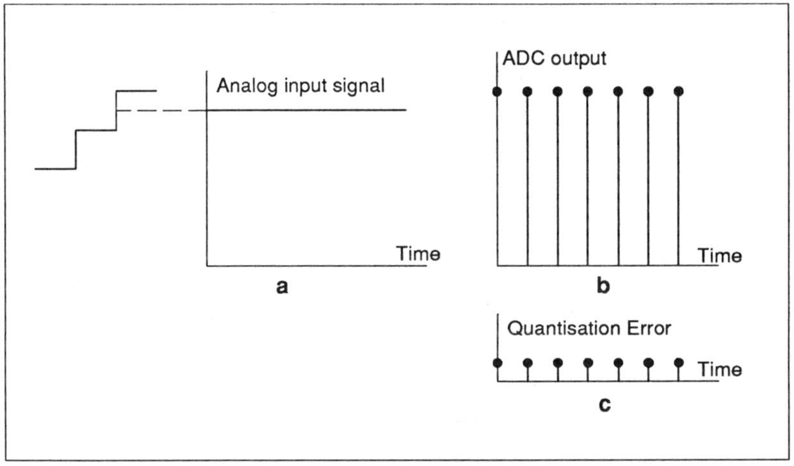

Fig. 3.13: The behaviour of a zero-order modulator with a DC input: a) the input, b) the quantised output, c) the error

In order to make this technique useful, it has to be ensured that

1) Only a small part of the quantisation noise is at low frequencies, no matter what the input signal is.

2) A more advantageous trade-off between the internal sampling rate f_s and the accuracy is required.

3.3.b. A Sigma-Delta modulator

A better approach to increase the ADC word length is depicted in Fig. 3.14.a. In this circuit, feedback is applied to suppress the quantiser nonlinearities. As in the previous circuit, the internal sampling frequency f_s is larger than the Nyquist rate: the OR defined in expression (3.1) is larger than one. The ADC output is fed back via an 11 bit DAC and compared with the analog input. The difference is applied to the ADC via a discrete-time lowpass filter H(z). For frequencies in the signal band, the filter gain is large to ensure a large loop gain. This circuit is a **Sigma-Delta modulator**.

It can easily be seen that the trade off between internal speed and accuracy is better for this circuit: when the average gain of the ADC is denoted by k_1 (with dimension [Volt^{-1}]) as depicted in Fig. 3.9.a, the ADC output equals

$$y(t) = k_1.v(t) + e(t) \tag{3.11}$$

where e(t) is the quantisation error. The DAC output can be denoted as

$$w(t) \; = \; k_2.y(t) \; = \; k_2.k_1.v(t) \; + \; k_2.e(t) \tag{3.12}$$

where k_2 stands for the DAC gain (dimension [Volt]). In the z-domain, the output $Y(z)$ can be calculated as:

$$Y(z) \; = \; \frac{k_1.H(z)}{1+k_1.k_2.H(z)}.X(z) \; + \; \frac{1}{1+k_1.k_2.H(z)}.E(z) \tag{3.13}$$

where $H(z)$ is the filter transfer function. $X(z)$, $Y(z)$ and $E(z)$ are the z-transforms of $x(t)$, $y(t)$ and $e(t)$ respectively. When the filter gain is large for frequencies in the signal band, the denominator of the second term of expression (3.13) is large and the in-band noise is suppressed. This is the main advantage of feedback: the quantisation noise is not only distributed over a wide frequency range as with the zero-order modulator, but in addition, the in-band noise is suppressed as shown in Fig. 3.14.b. As a result, $y(t)$ contains less in-band noise power and the SNR after the lowpass filter is higher. Depending on the **order** of the lowpass filter, Sigma-delta modulators are classified as First-order, Second-order etc. The circuit of Fig. 3.12. is sometimes denoted as a **zero-order sigma-delta modulator**.

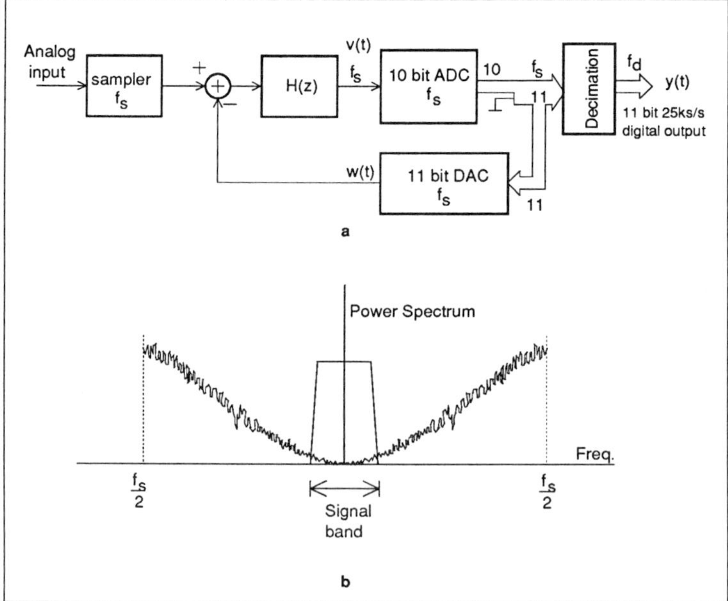

Fig. 3.14: a) the schematic of a Sigma-Delta modulator
b) the spectrum of y(t)

At this point, a special case is considered: suppose that the loop filter H(z) is an integrator with transfer function [31]

$$H(z) = \frac{1}{z-1} \tag{3.14}$$

Expression (3.13) becomes

$$Y(z) = \frac{k_1}{z-1+k_1.k_2}.X(z) + \frac{z-1}{z-1+k_1.k_2}.E(z) \tag{3.15}$$

In order to ensure stability, the pole of (3.15) must be situated within the z-plane unity gain circle, thus the product $k_1.k_2$ has to be between zero and two. Optimum stability is obtained when the pole is situated at the z-plane origin, thus when $k_1.k_2$ equals one. Then (3.15) yields:

$$Y(z) = \frac{1}{k_2.z}.X(z) + \frac{z-1}{z}.E(z) \tag{3.16}$$

The output equals the input delayed over one clock cycle, plus the first difference of the quantisation noise. The **noise transfer function** (NTF) (z-1)/z is depicted in Fig. 3.15. As can be seen, the noise at low frequencies is suppressed by feedback. This technique of transforming the noise spectrum is called **noise shaping**.

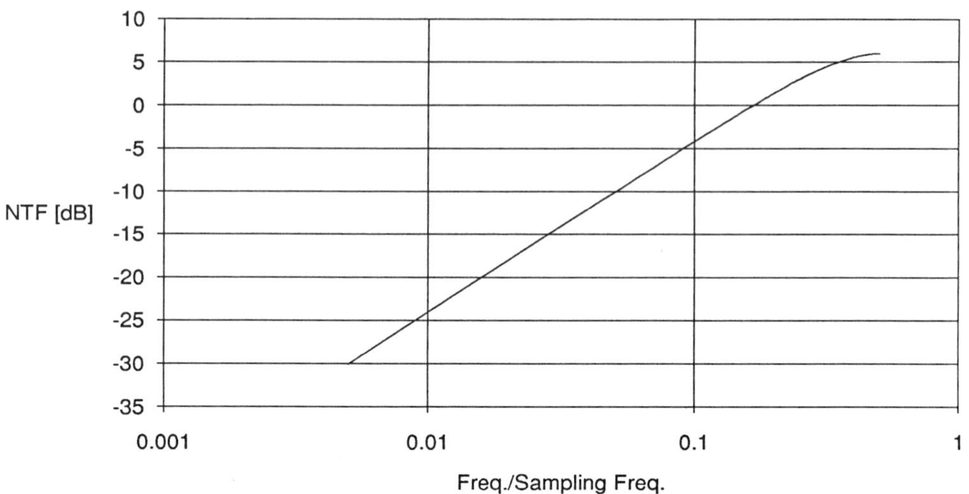

Fig. 3.15: Illustration of the noise shaping with a First-order modulator

To illustrate the principle further, let us consider the following example: Suppose the ADC in Fig. 3.14. is a 10 bit ADC for which the output levels are integer, between 0 and 1023. When a DC signal equal to 507.4 times V_{ref} is applied to the input, an output word equal to 507.4 would be exact. The output y(t) will oscillate between 507 and 508 as depicted in Fig. 3.16. An output equal to 507 appears more often than a 508 output, in such a way that the average output equals the input value of 507.4. The difference between output and input changes rapidly between -0.4 and +0.5. Hence, it can be understood intuitively that most of the quantisation noise spectrum power is at high frequencies. Fig. 3.16. shows that the requested output value of 507.4 is formed by interpolation between 507 and 508. This explains why oversampled data converters are sometimes denoted as **interpolating converters**. The **number** of 507's and 508's determines the average output value (507.4 here) while the **order** determines the error spectrum. In the same way as for the circuit of Fig. 3.12, a decimator can remove the outband noise and decrease the clock frequency, resulting in a larger output word at a lower clock rate.

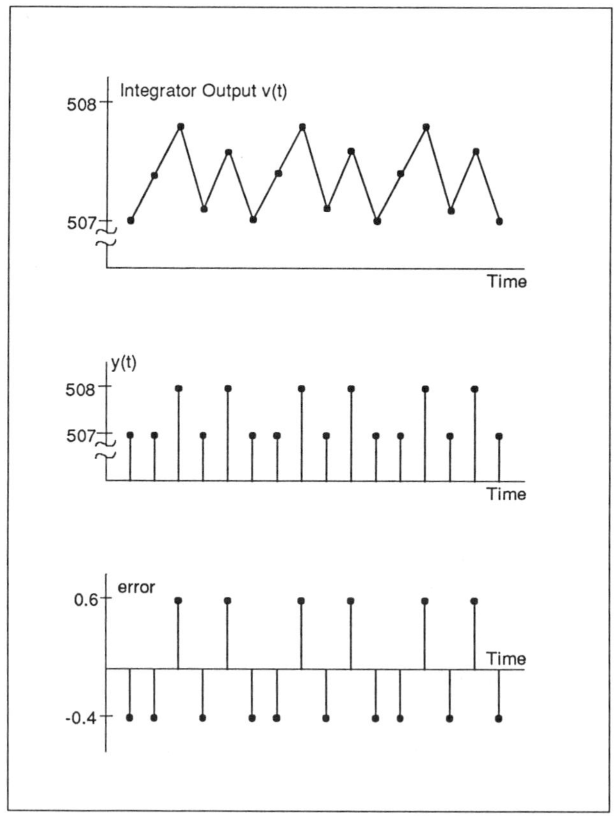

Fig. 3.16: Behaviour of a First-order Sigma-Delta
modulator with a DC input signal equal to 507.4

When the quantisation noise is assumed to be white and not correlated with the input signal, the SNR of the First-order modulator of Fig. 3.14 can be calculated [40]. The quantisation noise power spectrum is given by (3.16) where E(f) is expressed by (3.10). Provided that the lowpass filter at Fig. 3.14. is indefinitely steep, the noise power at the output equals:

$$N = \int_0^{f_{LPF}} \frac{1/3.[0.5\ \mathrm{LSB}]^2}{f_s/2} \, [z-1]^2 \Big|_{z=\exp(j2\pi f/f_s)} \, df \qquad (3.17)$$

with $f_{LPF} = f_d/2$

With a maximum signal power of $[2^B\ \mathrm{LSB}]^2/8$ and when the exponential is approximated by a Taylor series truncated after the linear term, this expression yields

$$\text{SNR} \approx 20.\log[2^B . \sqrt{1.5\pi} \ .\sqrt{3}. \ (\frac{\text{OR}}{\pi})^{3/2} \ ] \ \text{dB} \tag{3.18.a}$$

where $\text{OR} = f_s/f_d$ and $f_{LPF} = f_d/2$

B is the number of bits in the ADC output word. Expression (3.17) can be rewritten as

$$\text{SNR} \approx B.6 \ \text{dB} + r.9 \ \text{dB} - 3.4 \ \text{dB} \tag{3.18.b}$$

$$\text{where} \ \text{OR} = 2^r$$

i.e. doubling the internal sampling frequency (doubling OR) improves the SNR with 9 dB or 1.5 bit. Hence, the first drawback of the zero-order modulator is solved.

At the discussion of Fig. 3.13, it is mentioned that the quantisation noise is not white but correlated with the input signal. As can be seen from Fig. 3.16, the ADC input v(t) is not constant, even for a DC input signal. Therefore, consecutive samples of the quantisation noise of the circuit of Fig. 3.14 are less correlated than for the circuit of Fig. 3.12 and the noise spectrum E(z) is more white. But still, a Sigma-Delta modulator is a deterministic circuit and the **white noise assumption** is only an approximation. Expressions (3.16) to (3.18) are not exact. In Section 3.4, the consequences of this assumption will be investigated.

According to expression (3.13), the quantisation noise appears at the output after division by the loop gain. This principle can be further exploited by increasing the filter gain H(z) in the signal band. For instance when H(z) is a series of two integrators as depicted in Fig. 3.17, the output equals [32-33]:

$$Y(z) = \frac{k_1}{(z-1)^2 + k_1.k_2.b.(z-1) + k_1.k_2}.X(z)$$

$$+ \ \frac{(z-1)^2}{(z-1)^2 + k_1.k_2.b.(z-1) + k_1.k_2}.E(z) \tag{3.19}$$

Without the dashed branch in Fig. 3.17 (thus b equal to zero), the z-domain poles are situated outside the unity gain circle. Optimum stability is obtained with $k_1.k_2$ equal to one and b equal to two. Under these conditions, expression (3.19) yields:

$$Y(z) = \frac{1}{k_2}.\frac{1}{z^2}.X(z) + \frac{(z-1)^2}{z^2}.E(z) \tag{3.20}$$

The output equals the input, delayed over two clock cycles plus the second difference of the quantisation noise. The noise transfer function $(z-1)^2/z^2$ is plotted in Fig. 3.18. Compared with Fig. 3.15, the low frequency noise is further suppressed while the out-band noise is enhanced. The in-band noise power can be calculated from:

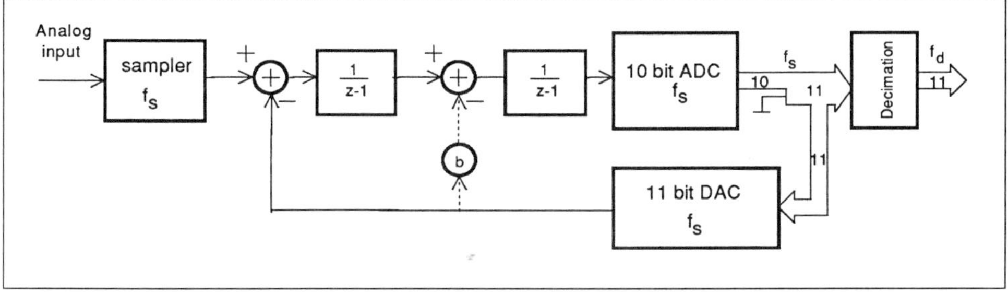

Fig. 3.17: A Second-Order Sigma-Delta modulator

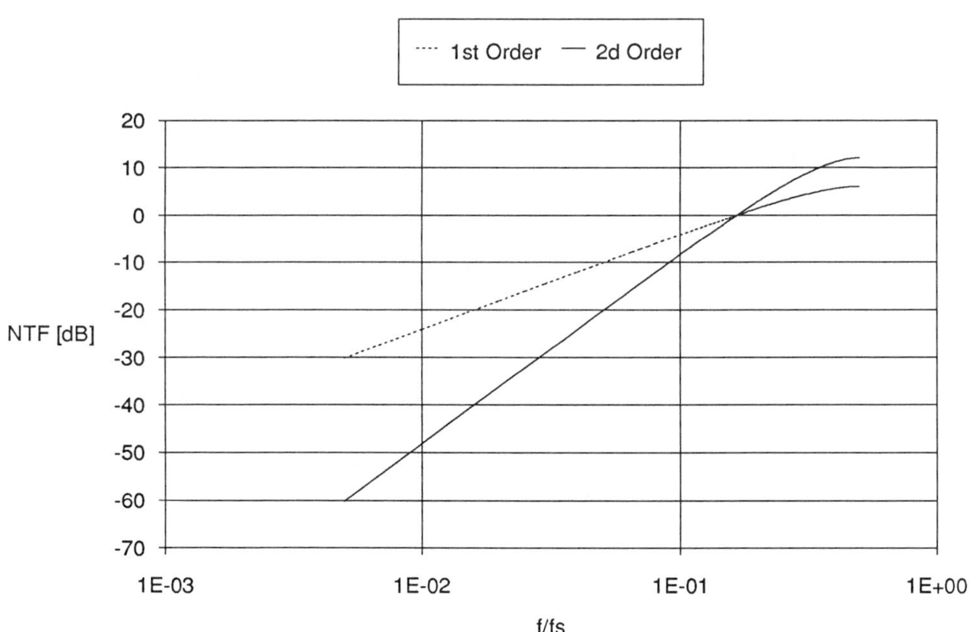

*Fig. 3.18: The Noise transfer function of a Second-order
modulator*

$$N = \int_0^{f_{LPF}} \frac{1/3 \cdot [0.5 \text{ LSB}]^2}{f_s/2} \, [e^{j2\pi f/f_s} - 1]^4 \, df \qquad (3.21)$$

and

$$\text{SNR} \approx 20 \cdot \log[2^B \cdot \sqrt{1.5\pi} \cdot \sqrt{5} \cdot \left(\frac{\text{OR}}{\pi}\right)^{5/2}] \text{ dB} \qquad (3.22)$$

$$= B.6 \text{ dB} + r.15 \text{ dB} - 11 \text{ dB} \qquad (3.23)$$

where $\text{OR} = 2^r$ and B is the output word length

Each doubling of the internal sampling frequency yields 2.5 bit of accuracy improvement.

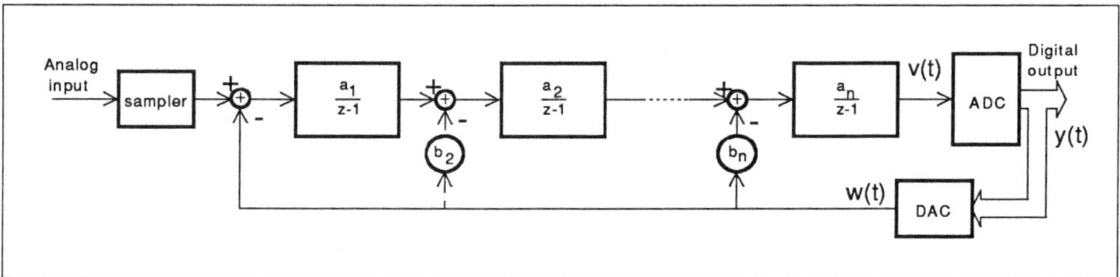

Fig. 3.19: a n-th order Sigma-Delta modulator

In general, with an n-th order filter as depicted in Fig. 3.19, with a proper choice of the coefficients a_i and b_i, the poles can be situated at the z-plane center and the output becomes

$$Y(z) = \frac{1}{k_2 \cdot z^n} \cdot X(z) + \frac{(z-1)^n}{z^n} \cdot E(z) \qquad (3.24)$$

The in-band noise power equals

$$N = \int_0^{f_{LPF}} \frac{1/3 \cdot [0.5 \text{ LSB}]^2}{f_s/2} \, [e^{j2\pi f/f_s} - 1]^{2n} \, df \qquad (3.25)$$

which yields

$$\text{SNR} \approx 20 \cdot \log[2^B \cdot \sqrt{1.5\pi} \cdot \sqrt{2n+1} \cdot \left(\frac{\text{OR}}{\pi}\right)^{(2n+1)/2}] \text{ dB} \qquad (3.26)$$

The resolution increases with n+0.5 bit per octave of oversampling ratio. The number of integrators n is denoted as the **order** of the Sigma-Delta modulator.

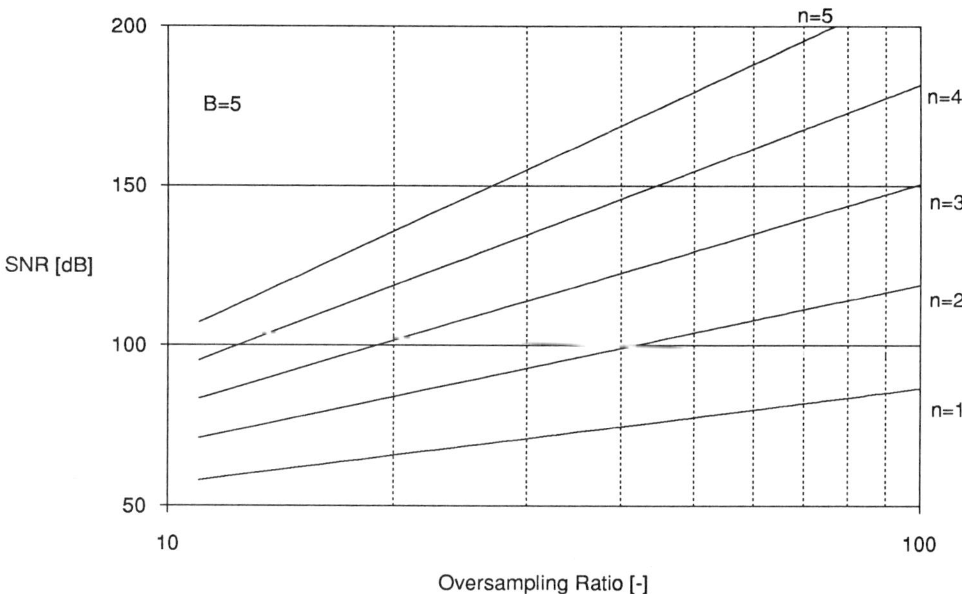

Fig. 3.20: SNR versus Oversampling Ratio for different
orders of the loop filter

In Fig. 3.20, the SNR according to expression (3.26) is plotted for an ADC with B equal to 5 bit and for different orders n of the filter. The higher the order, the lower the requested oversampling ratio for a given resolution and the lower the required internal sampling frequency f_s. This is the first benefit of increasing the order.

3.3.c. A Single-bit Sigma-Delta modulator

Up till now, it is assumed that the DAC in Fig. 3.14, Fig. 3.17. and Fig. 3.19. is ideal, i.e. that the output levels have the exact value and the bins all have equal size. Since the DAC is in the feedback path, its nonlinearities are not suppressed by feedback. Hence, although for instance a fast 10 bit ADC can be applied to construct a slower 11 bit ADC, the accuracy of the DAC in the feedback path has to be 11 bit. The problem is shifted to the resolution of the DAC.

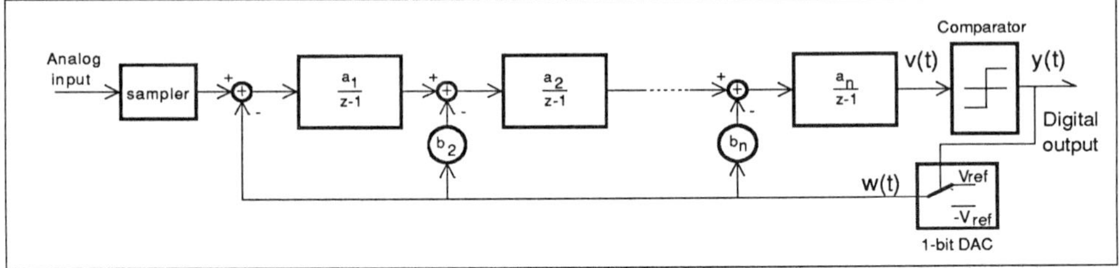

Fig. 3.21: A Single-bit Sigma-Delta modulator

A solution to overcome this is to use a one bit ADC and DAC as depicted in Fig. 3.21. Since a single bit DAC has only two output levels, there is only one bin and no nonlinearity problem. The SNR is given by [33]:

$$SNR \approx 10.\log\left[6\pi \;.2n{+}1.\left(\frac{OR}{\pi}\right)^{2n+1}\right] \; dB \qquad\qquad (3.27)$$

where n is the order. This relationship is depicted in Fig. 3.22. For example for a second order modulator, an oversampling ratio of 50 yields 80 dB or 13 bit. The same performance can be obtained with a fourth-order modulator with an oversampling ratio of 15. A higher resolution can be obtained with a higher oversampling factor. A higher OR or a higher order yield a higher resolution. However, we remind the reader that this expression (3.27) is based on the assumptions that the quantisation noise e(t) is white and that the poles in (3.13) are situated at the z-plane origine. In sections 3.3.d. and 3.4, it will be pointed out that these assumptions are not valid. In reality, the SNR predicted by (3.27) can be too optimistic.

The quantiser output of Fig. 3.21. oscillates between one and minus one and the input signal is represented by the density of ones. This signal is a **pulse density modulation (PDM)** signal.

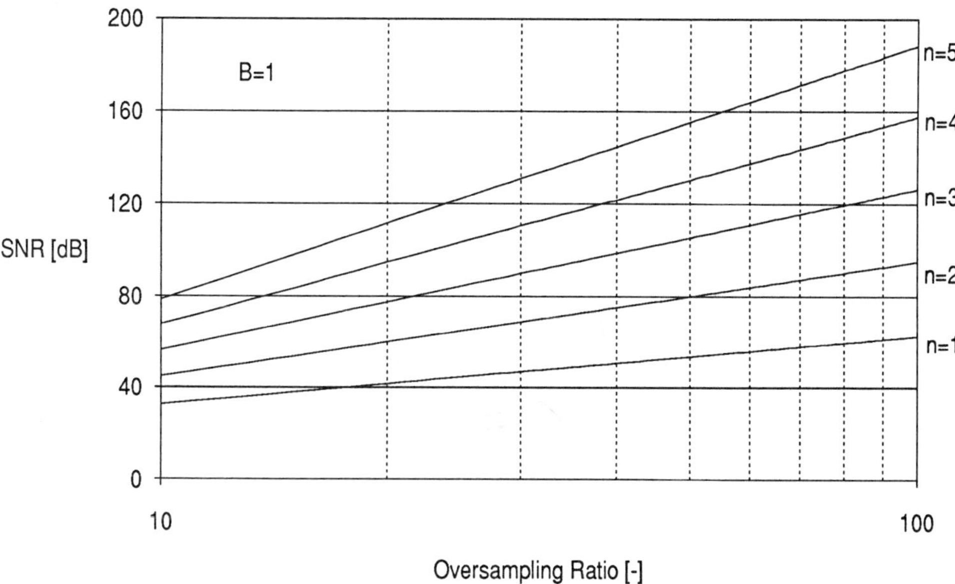

*Fig. 3.22: SNR versus Oversampling ratio according to
expression (3.27)*

Fig. 3.23.a. shows a complete single-bit Sigma-Delta ADC. The input signal passes an **anti-aliasing** filter to remove spurious signals above half of the sampling frequency f_s and to ensure that the signal is band limited. The loop filter H(z) is a (switched-capacitor) discrete-time analog lowpass filter that performs the sampling and the filtering. The one-bit ADC is a comparator and the DAC is a switch between two reference voltages. The one-bit digital output is then decimated [34-36] to obtain a larger digital output word at a lower rate f_d.

In Fig. 3.23.b, the same principle is applied to construct a DAC. The digital word is upsampled from the sampling frequency f_d to f_s. The Sigma-Delta modulator is fully digital: the filter is digital, the "comparator" output is just the inverse of the sign bit and the "DAC" is a digital code conversion. The resulting one bit signal is converted to an analog quantity and lowpassed in an analog smoothing filter, named a **reconstruction filter**.

In Fig. 3.23.c, the Sigma-Delta principle is used in its earliest application: a word length reduction for digital data transmission. At the sender, the word length is reduced to B_2 (usually one) bit by a digital Sigma-Delta modulator. At the receiver, the signal is digitally decimated to reconstruct the original signal.

In the three cases, the circuit is extremely tolerant to process characteristics: although the resolution can be high, no precise analog components are required.

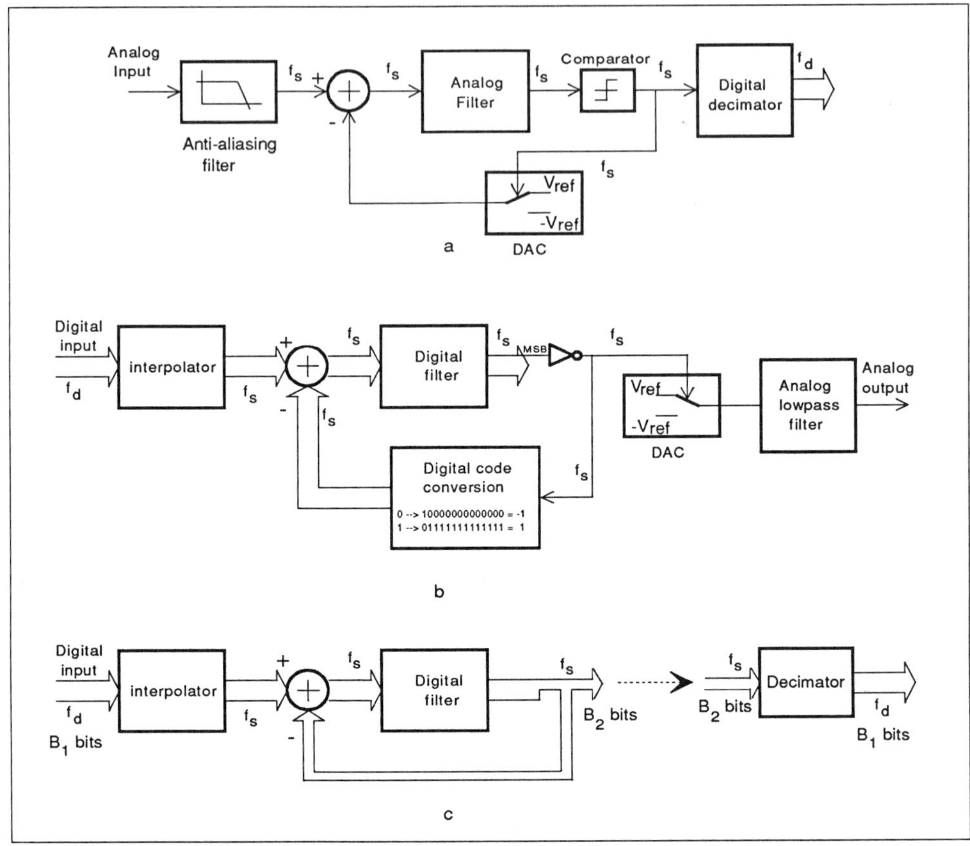

Fig. 3.23: Principles of a) a Sigma-Delta ADC, b) a Sigma-Delta DAC and c) a word length reduction

3.3.d. Stability problems in Single-bit Sigma-Delta modulators

Up till now, the poles of the Sigma-Delta modulator loop are calculated with a linearised ADC. The linear gain is denoted as k_1 in Fig. 3.9.a. and the quantisation error is assumed between -0.5 LSB and +0.5 LSB. For a multi-bit quantiser, this is a reasonable approximation provided that the quantiser is not **overloaded**. Thus v(t) has to be between -2^{B-1}-0.5 bin and 2^{B-1}+0.5 bin in Fig. 3.9.a. According to expression (3.24), the quantiser input voltage v(t) of an n-th order multi-bit Sigma-Delta modulator equals

$$v(t) = x(t-nT) + \{ \nabla^n e(t) - e(t) \}/k_1 \qquad (3.28)$$

where ∇^n stands for the n-th backward difference. Since e(t) is always between -0.5 LSB and +0.5 LSB, ∇e(t) is between -1 LSB and +1 LSB, ∇^2e(t) is between -2 LSB and +2 LSB and ∇^ne(t) is between -2^{n-1} LSB and $+2^{n-1}$ LSB. The quantiser is not overloaded if

$$| x(t) | \; < \; 2^{B-1}+0.5 \; - \; \{ \; 2^{n-1}-0.5 \; \} \quad \text{bin} \tag{3.29}$$

This condition is depicted in Fig. 3.24: the input amplitude has to be smaller than the ADC input range.

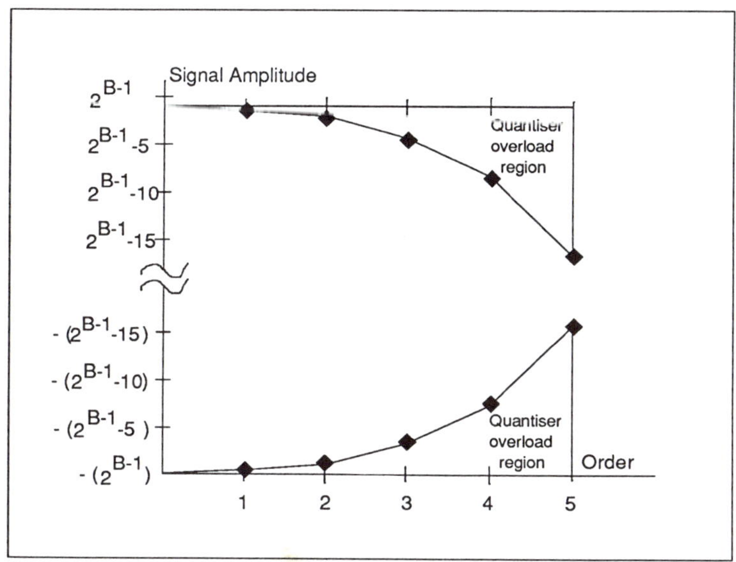

Fig. 3.24: Input amplitude reduction for stability requirements

When condition (3.29) is satisfied, a multi-bit quantiser can be approximated by a **linear gain** k_1 and a noise source (see Fig. 3.9.d.). The output then equals (see Fig. 3.19.)

$$Y(z) = \frac{a_1 a_2 a_3 \ldots a_n . k_1 . X(z) \; + \; (z-1)^n . E(z)}{(z-1)^n \; + \; k_1 k_2 b_n a_n (z-1)^{n-1} + \; k_1 k_2 b_{n-1} a_n a_{n-1} (z-1)^{n-2} + k_1 k_2 b_{n-2} a_n a_{n-1} a_{n-2} (z-1)^{n-3} + \; \ldots \; + \; k_1 k_2 a_n a_{n-1} a_{n-2} \ldots a_1} \tag{3.30}$$

With linear circuit theory techniques, the poles of (3.30) can be calculated as demonstrated for the expressions (3.15) and (3.19). By a proper choice of the filter coefficients a_i and b_i, the poles can be situated within the z-plane unity gain circle. Note that the locations of the poles are functions of the quantiser gain k_1. The quantisation error will cause low-level limit cycles, equivalent to the limit cycles caused by the truncation in a digital filter.

For a single bit quantiser on the other hand, the linearisation of the quantiser is a rough approximation, as illustrated in Fig. 3.25: three linear approximations are shown and there is no reason to prefer one over the other. Therefore, it is not clear which value for k_1 to take for the stability analysis. The one-bit truncation by the quantiser can cause high-level limit cycles which will cause an overload of the modulator lowpass filter, resulting in a clipped filter output signal. Hence, the Sigma-Delta modulator will not operate properly.

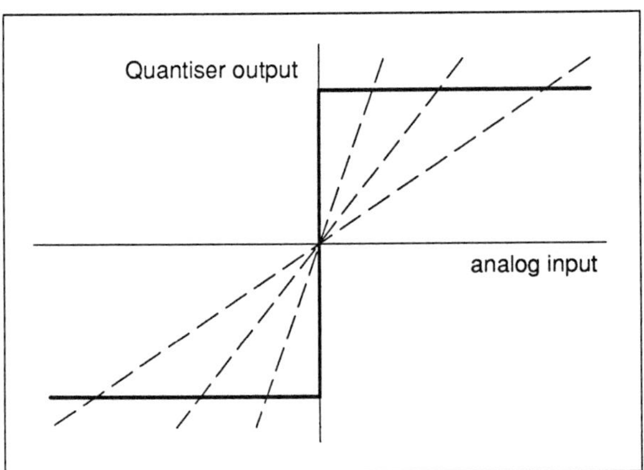

Fig. 3.25: Three linear approximations of a Single-bit quantiser

In the literature, several techniques are proposed to linearise the quantiser nonlinearity and to obtain simple stability conditions and SNR expressions. In the popular work of Agrawal and Shenoi [31], a stability analysis is performed where the problem of the unknown k_1 is neglected and the arbitrary value of one is assumed. The results obtained in this work are in disagreement with the reality. For instance, according to expression (3.30) the locations of the poles are functions of the last integrator gain a_n. In reality however, this gain is arbitrary since the comparator is sensitive to the sign only and not to the magnitude of v(t). When varying a_n, only the magnitude and not the sign of v(t) is changed. Therefore, varying a_n causes no change of the circuit behaviour.

In previous attempts to derive mathematical stability conditions for a one-bit Sigma-Delta modulator loop [37], [39], [40], the quantiser is modeled by a linear, signal-dependent gain determined from a least-square fitting. This is the "describing function technique" [38], well known in nonlinear system analysis. However, a basic condition for the validity of this technique is that for a sinusoidal input voltage, v(t) is approximately sinusoidal, i.e. that the quantisation noise at the comparator input is small. For a Single bit Sigma-Delta modulator, this condition is not met.

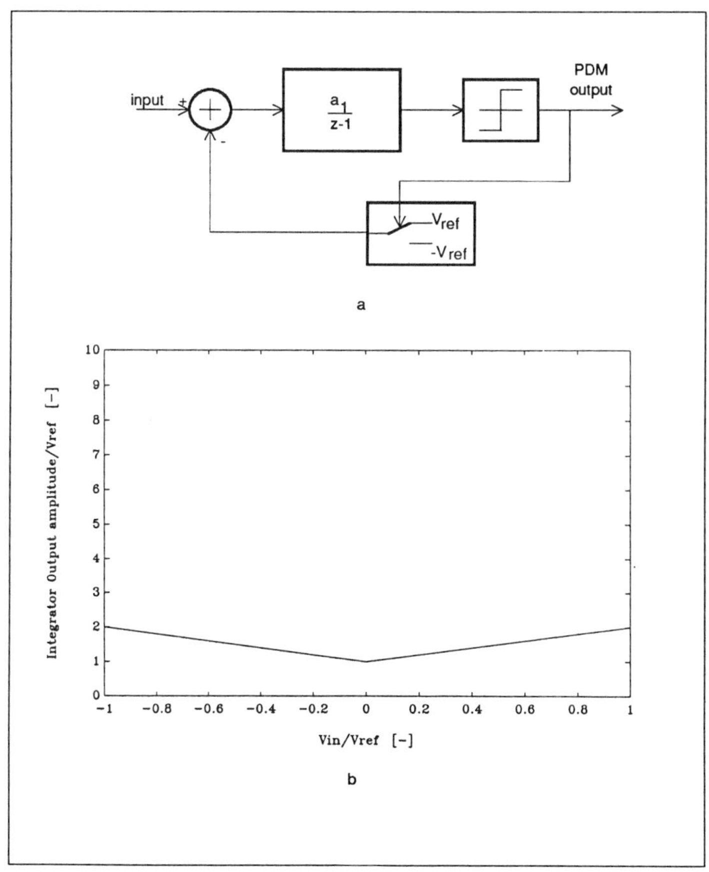

Fig. 3.26: Stability of the First-order modulator:
a) schematic
b) integrator output amplitude for a_1 equal to one.

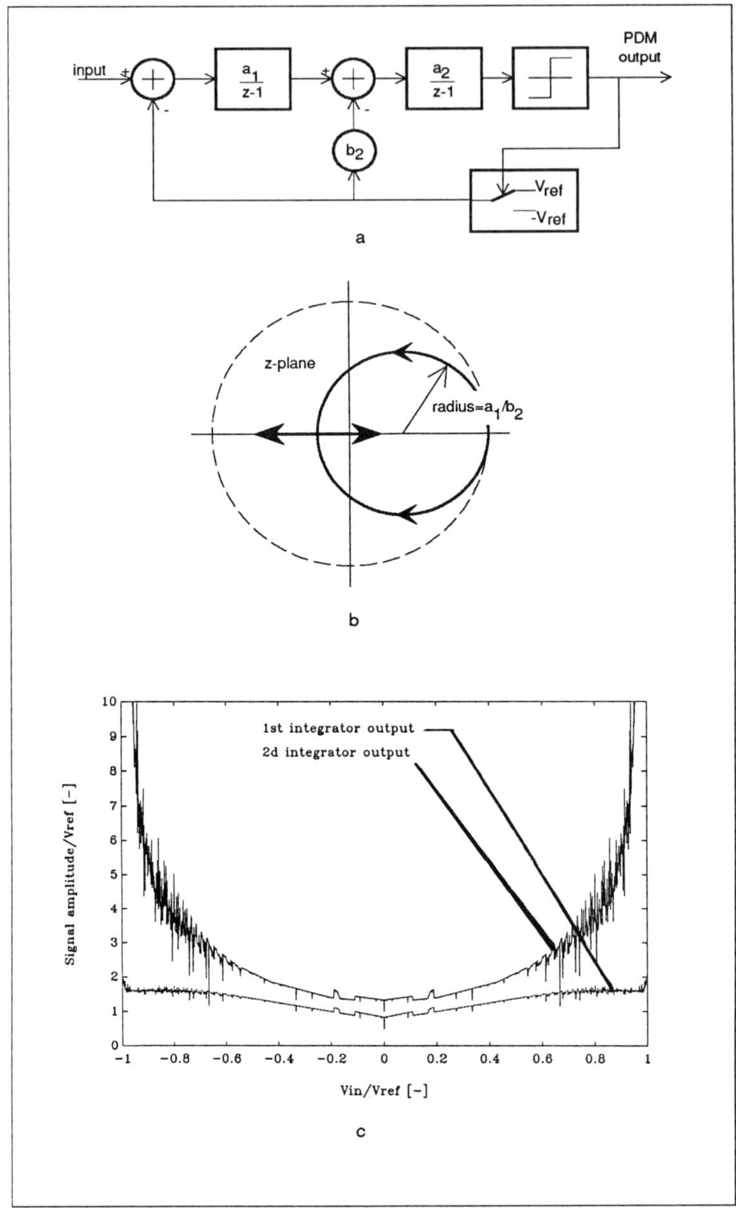

Fig. 3.27: The stability of a Second-order modulator
a) schematic
b) root locus for varying k_1 and
c) the integrator output amplitudes versus the input signal
for $a_1=a_2=1$ and $b_2=2$

In chapter 4, section 2, it will be shown that the *First-order* Single-bit Sigma-Delta modulator of Fig. 3.26.a. is stable for input values x(t) between $-V_{ref}$ and $+V_{ref}$, independent of the integrator gain a_1. For a_1 equal to unity, the amplitude at the integrator output is a function of the input signal as depicted in Fig. 3.26.b. and is always between $-2.V_{ref}$ and $+2.V_{ref}$. A scaling of this output range can be performed by changing the integrator gain a_1.

In order to investigate the stability of the *Second-order* modulator loop op Fig. 3.27.a, the following intuitive approach can be used [40]. When linearising the quantiser as shown in Fig. 3.25, a linear system is obtained. The two modulator loop poles can be calculated from expression (3.30). When these poles are inside the z-plane unity-gain circle, the loop is stable. However, as illustrated in Fig. 3.25, it is not clear which value for k_1 to take to calculate the poles. In Fig. 3.27.b, a root locus of the poles is drawn with k_1 as a parameter. The location of the locus is a function of the parameters a_1 and b_2 only. The root locus is entirely inside the unity gain circle when

$$b_2 > a_1 \qquad\qquad (3.31)$$

Under this condition, the modulator is stable, regardless of the quantiser gain value. Hence, an accurate determination of the quantiser gain is of no importance.

This intuitive approach will be verified in chapter 4. It will be shown that condition (3.31) is not completely in agreement with the reality: stability requires a value for b_2 larger than 1.25 times a_1 [32].

The integrator output amplitudes are plotted in Fig. 3.27.c. versus the input voltage x(t). As can be seen, the second integrator output $v_2(t)$ increases without limit when the input signal amplitude approaches V_{ref}. This indicates that for large input signals, the loop damping is low, resulting in a huge internal nonlinear resonance. As a result, the second integrator will overload for a too large input signal and the Second-order Sigma-Delta modulator loop will not operate properly. This explains the SNR degradation for large input values which is noticed in most Sigma-Delta modulators. An example of this SNR degradation will be given further on in Fig. 3.31.

In Fig. 3.28.b, the root locus for the *Third-order* modulator of Fig. 3.28.a. is depicted. For the quantiser gain k_1 equal to zero, the three poles are situated at z equal to one. For increasing k_1, two branches start with an angle of 60° from this point, regardless the a and b values of Fig. 3.28.a. As a result, for small k_1, two poles will be outside the unity gain circle and the loop will be unstable. Hence, for a third (or higher) order modulator a stability condition similar to (3.31) valid for all k_1 values cannot be obtained.

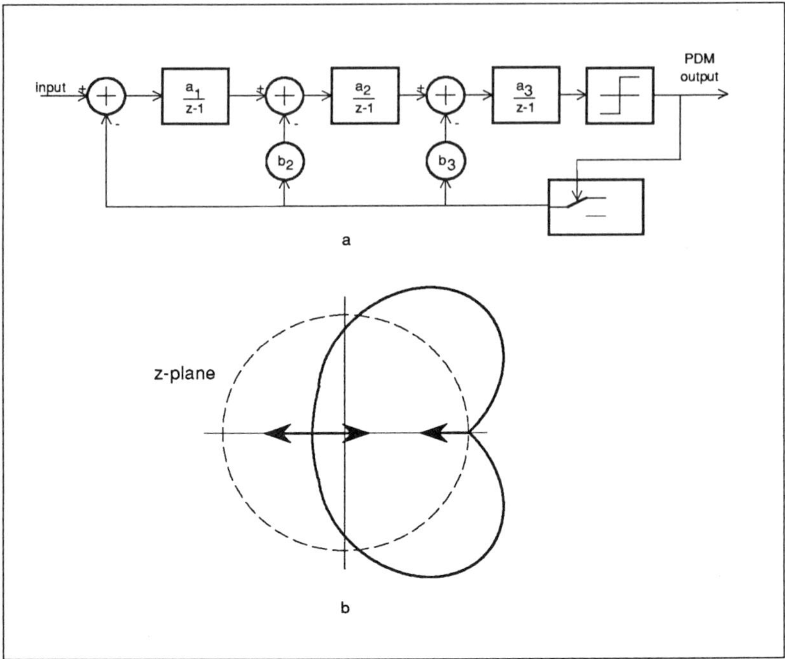

Fig. 3.28: The stability of a Third-order modulator loop a)
schematic b) root locus

From the previous discussion, it can be concluded that First- and Second-order Sigma-Delta modulators can be realised without stability problems. Higher-order modulator loops are conditionally stable: special measures have to be taken to prevent a low-frequency nonlinear oscillation or an overload of the integrators. The discussion of these techniques is beyond the scope of the general introduction presented in this chapter but will be treated in detail in Chapter 4.

3.4. THE QUANTISATION NOISE OF SIGMA-DELTA MODULATORS

At the discussion of the zero-order Sigma-Delta modulator, it was pointed out that this circuit had two drawbacks: the resolution-for-speed trade-off is not very advantageous and the quantisation noise is not white. For Sigma-delta modulators with a higher order, the resolution-for-speed trade-off is more advantageous, but the quantisation noise is still not white and depends upon the input signal.

In this section, two situations are discussed where the effects due to the colored quantisation noise become predominant [41-44]. First, for a DC input signal, the quantisation noise spectrum consists of discrete spectral lines, called **pattern noise**. This noise spectrum can result in an SNR which is much worse than the SNR predicted by expression (3.27). And second, for a sinusoidal input signal with a low signal amplitude, the nonlinear circuit behaviour an cause an output signal that does not even contain the input signal component. This effect is further described as **Low signal-level Distortion**.

3.4.a. The quantisation noise of a First-order Sigma-Delta modulator driven by a DC input signal

Consider the First-order Single-bit Sigma-Delta modulator of Fig. 3.29.a. The nonlinear differential equations describing this integrator type can be solved exactly [41]. This allows us to study this modulator type more in detail, by analytical means.

Since the integrator gain is infinite for a DC signal, the feedback loop forces the DC component at the integrator input to equal zero. Hence the DC component of y(t) equals the DC component of x(t) and the quantisation noise contains no DC component.

When the input voltage x(t) is constant, equal to zero, it can be easily verified by hand that the output y(t) oscillates between 1 and -1 as depicted in Fig. 3.29.b. Its average equals the input. The output spectrum contains energy at half of the sampling frequency only. The quantisation noise spectrum is highly colored, but its power spectrum is entirely outside the signal band. This illustrates that the white noise assumption is not valid. Fortunately, the in-band noise is not degraded by the noise peak.

Consider now the situation where the input is a small DC value, equal to

$$x(t) \ = \ \frac{1}{2p+1} \tag{3.32}$$

where x is the analog input, scaled to V_{ref} and p is a large positive integer. The output is depicted in Fig. 3.29.c: y(t) oscillates between one and minus one but every 2p+1 samples, there is an extra one. For a negative x, the output is similar but of opposite sign. The quantisation noise spectrum is colored, with discrete lines given by:

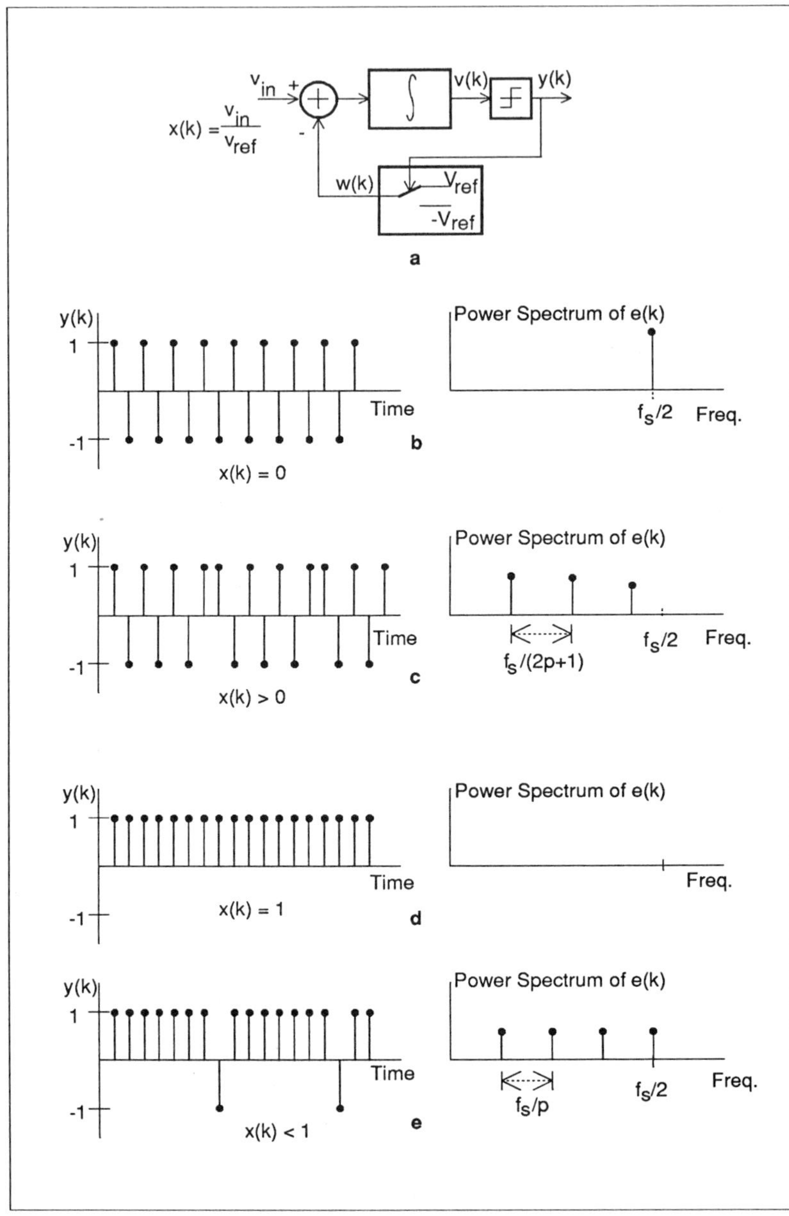

Fig. 3.29: a) a First-order Sigma-Delta modulator b) The output for x=0 c) the output for x=1/7 d) the output for x=1 e) the output for x=3/4

$$E^2(f) \;=\; \frac{2.x^2}{\cos^2(\pi.f/f_s)} \qquad \text{at frequencies given by} \qquad f=k.|x|.f_s \quad k>0$$

$$(3.33)$$

with k integer and positive.

This power spectrum is colored but now there are spectral lines inside the signal band. This degrades the SNR. Since the signal band is much smaller than the sampling frequency, (3.33) approximately yields

$$E^2(f) \;=\; 2.x^2 \qquad \text{for f in the signal band} \qquad (3.34)$$

Since the number of in-band lines is given by

$$\frac{\text{signal-band}}{|x|.f_s} \;=\; \frac{f_s/(2OR)}{|x|.f_s} \qquad (3.35)$$

the total in-band noise power equals

$$N \;=\; |x|/OR \qquad (3.36)$$

as long as there is at least one in-band spectral line, i.e. as long as

$$|x| < 1/(2.OR) \qquad (3.37)$$

In Fig. 3.30.a, the in-band noise power is depicted versus x for an oversampling ratio of 64. The noise increases with increasing $|x|$ according to expression (3.36). With increasing $|x|$, the frequencies of the noise spectral lines vary according to expression (3.33) and move out of the signal band. When the last line moves out band, the quantisation noise decreases abruptly. The maximum in-band noise value is given by (3.36) when x equals (3.37):

$$N_{max} \;=\; 1/(2.OR^2) \qquad (3.38)$$

In Fig. 3.30.b, the quantisation noise is depicted for values of x, close to unity. (A similar calculation can be made for x close to minus one). Consider the case where the input signal equals V_{ref}, thus x equal to unity. The output signal is depicted in Fig. 3.30.d: y(t) is always equal to one and the quantisation noise is zero.

When the input x is a little smaller than one and equal to

$$x(t) \;=\; 1-2/p \qquad (3.39)$$

a

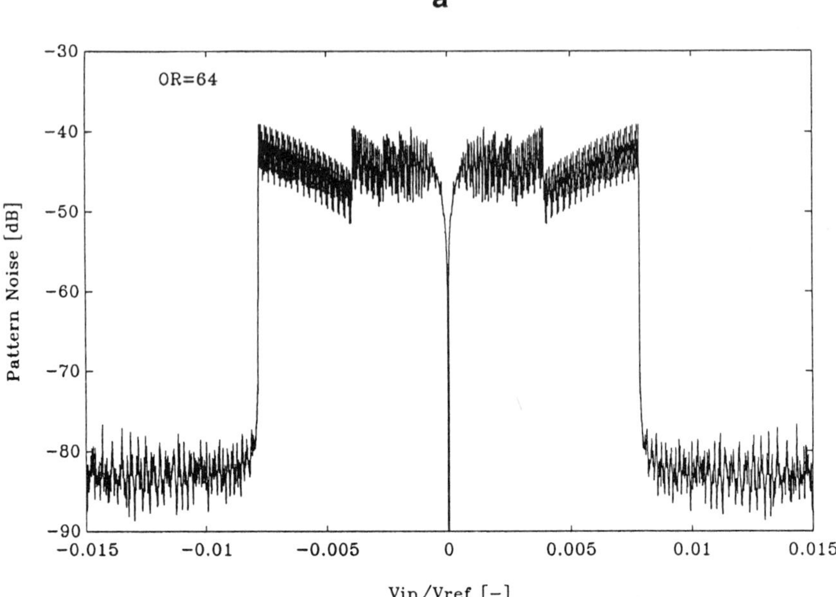

b

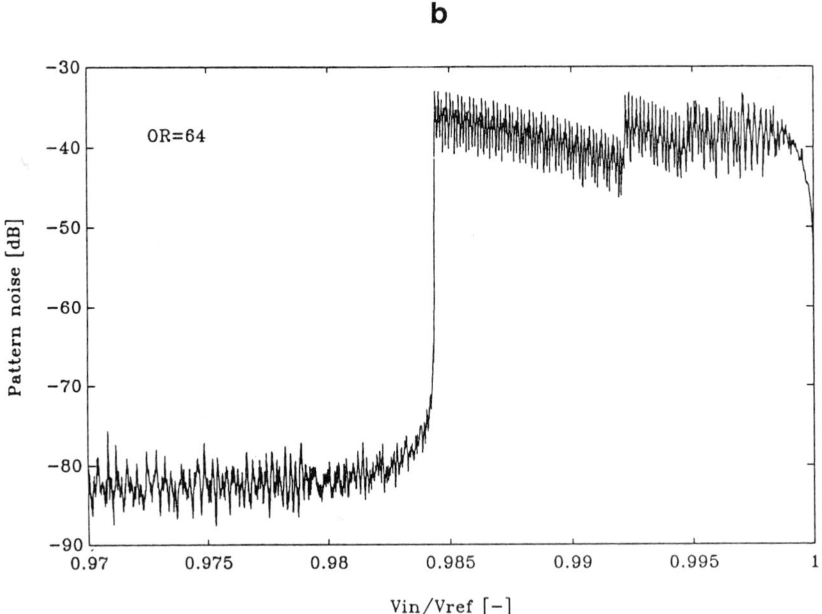

c

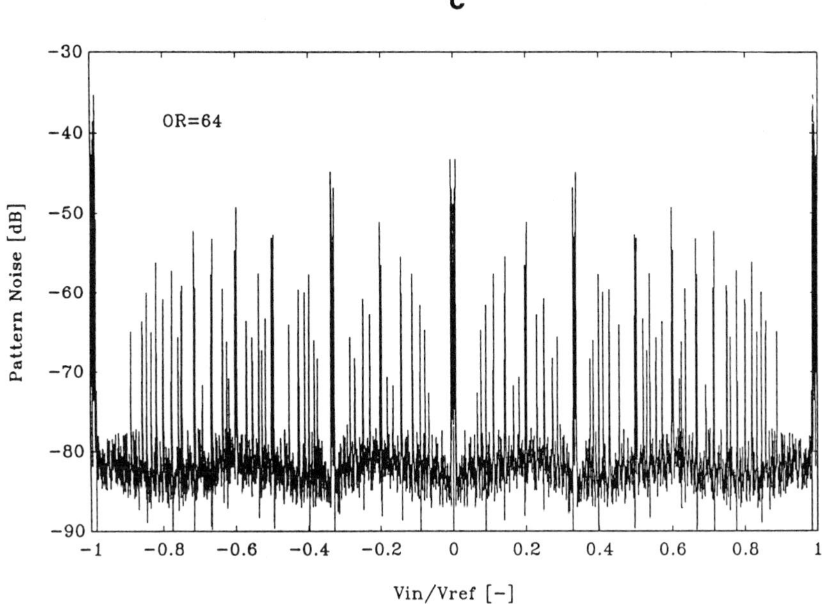

*Fig. 3.30: The in-band quantisation noise for a) x around
zero b) x approx. one and c) the total input range*

with p integer and positive, the output y(t) is as depicted in Fig. 3.29.e: one of every p
samples equals minus one. Again, the noise power spectrum is discrete with spectral
lines in the signal band:

$$E^2(f) = 2.(1-x)^2 \qquad \text{with } f = k.(1-x).f_s/2 \qquad k>0 \qquad (3.40)$$

with k integer. The number of in-band samples equals

$$\frac{signal-band}{(1-x).f_s/2} = \frac{f_s/(2.OR)}{(1-x).f_s/2} \qquad (3.41)$$

and the total in-band noise power is given by:

$$N = \frac{2.(1-x)}{OR} \qquad (3.42)$$

as long as there is at least one spectral line in the signal band, thus as long as

$$1-x < 1/OR \qquad (3.43)$$

The worst case in-band noise power is expressed by (3.42) with x given by (3.43):

$$N_{max} = 2/OR^2 \qquad (3.44)$$

In Fig. 3.30.c, the quantisation noise is plotted versus the DC input signal x(t). This drawing shows sharp peaks close to plus and minus one (expressed by (3.40) to (3.44)), around zero (see expressions (3.36) to (3.38)) and, in a similar way, around every rational number where the denominator is a small integer number. When x is rational, satisfying

$$|x| = 1 - \frac{2.q}{p} \qquad (3.45)$$

with p and q integer and positive and no common dividers in p and q, there are two quantisation noise peaks around x with a maximum value given by [42]

$$N_{max} = 2/(p.OR)^2 \qquad (3.46)$$

 E.g. for an oversampling ratio of 64, the noise peaks around x=2/3 (p=6 and q=1) are 15.5 dB below the peak at x=1.

Since the largest peaks are close to one and minus one, the maximum in-band noise is expressed by (3.44). E.g. for an OR of 100, the maximum in-band noise power is -37 dB. This value is much worse than the -65 dB obtained from expression (3.16). The extra in-band noise power due to the colored quantisation noise is called **pattern noise**. Increasing the OR with a factor of two yields one bit resolution.

3.4.b. The Quantisation noise of higher-order Sigma-Delta modulators for a DC input

A First-order Sigma-Delta modulator generates pattern noise. For the special case of a DC input, this noise is expressed by formulae (3.39) to (3.46). Basically, there are two solutions to eliminate this pattern noise:

- A first possibility is to deliberately add high-frequency noise (named **dither**) to the input signal [45]. When this noise contains no energy in the signal band, it will be removed afterwards by the decimating filter. Due to the dither, the input is no longer constant and the sharp peaks in the quantisation noise spectrum are smoothed out. However, the total in-band noise power is not substantially decreased.

- A better solution is to increase the order of the modulator. In Fig. 3.31, the quantisation noise of a Second-order Sigma-Delta modulator is plotted versus the input signal. The nonlinear differential equation describing this modulator type cannot be solved up till now. However, it can be verified experimentally that a

Second-order modulator is less sensitive to pattern noise. For inputs up to 0.9 times V_{ref}, the quantisation noise spectrum is approximately constant. For higher input signals, the last integrator will overload (see Fig. 3.27.b.) and the in-band noise increases.

In general, increasing the order of the modulator decreases the SNR degradation caused by pattern noise. This is a second advantage of higher-order modulators.

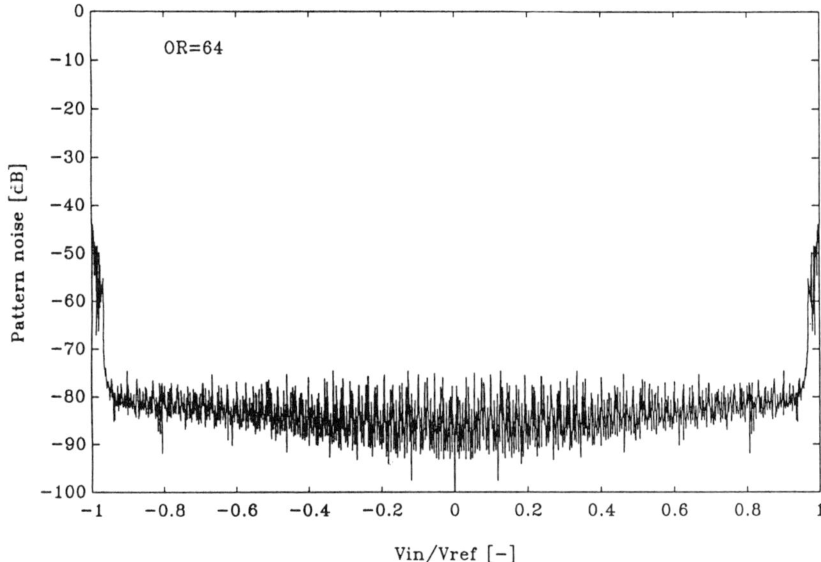

Fig. 3.31: The in-band Quantisation noise of a Second-order modulator

3.4.c. The quantisation noise for small sinusoidal input signals

When the input voltage of the Single-bit First-order modulator of Fig. 3.29.a. equals zero, the output y(t) oscillates between one and minus one as plotted in Fig. 3.32.a. The integrator output v(t) (Fig. 3.32.b.) oscillates between $a_1.V_{ref}$ and $-a_1.V_{ref}$.

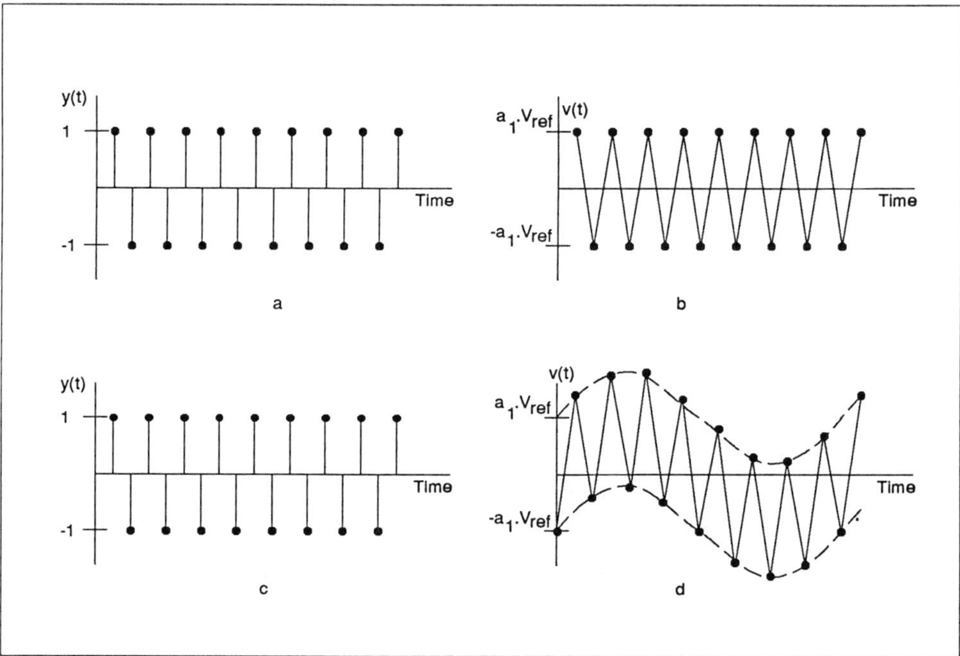

Fig. 3.32: The mechanism of Low signal-level Distortion

When a small sinusoidal input signal with amplitude V_{in} and frequency f is applied, the integrator output is as depicted in Fig. 3.32.d: a sinusoid with amplitude $V_{in}*|H(f)|$ is superimposed upon the waveform of Fig. 3.32.b. When the amplitude of this sinusoid is smaller than $a_1.V_{ref}$, the sign of the signals depicted in Fig. 3.32.b. and Fig. 3.32.d. are always the same. Since the comparator of Fig. 3.32.a. decides according to the sign of v(t) and not to the magnitude, the output y(t) is not changed. Hence, the small analog input voltage causes no change of the output. This indicates that the resolution of the Sigma-Delta modulator is limited [46].

To produce an output signal, the input signal amplitude has to satisfy

$$V_{in} \cdot \left| \frac{a_1}{z-1} \right| > a_1 \cdot V_{ref} \quad \text{with } z = \exp(j2\pi \cdot f/f_s) \qquad (3.47)$$

The worst case for this condition is when the input frequency is on the edge of the signal band. Then (3.47) becomes approximately:

$$V_{in}/V_{ref} > \pi/OR \qquad (3.48)$$

For a resolution of B bit, this treshold should be below 2^{-B}. Note that this condition is more severe than the SNR condition (3.27).

It can be concluded that the quantiser nonlinearity limits the resolution of the Sigma-Delta modulator. Input signals below this resolution are not reflected to the output. This resolution can be improved by increasing the OR or the modulator order. A doubling of OR yields only one bit of accuracy improvement (1.5 bit according to expression (3.27))

In a similar way, it can be pointed out that for a Second-order modulator, the resolution is given by

$$V_{in}/V_{ref} > 0.75 \ (\pi/OR)^2 \qquad (3.49)$$

Doubling the OR yields two bits accuracy (in stead of 2.5 according to (3.27)).

3.4.d. Conclusions

By linearising the comparator characteristic and by introducing the white noise assumption, expression (3.27) for the trade-off between oversampling ratio and resolution can be found. However, for some special input signals, the noise power predicted by this equation is far too optimistic:

- As explained in section 3.4.a, the quantisation noise for a First-order modulator driven by a DC input is not white and the in-band noise, expressed by (3.44), is larger than the value from (3.27). The problem can be solved by increasing the modulator order.
- The comparator linearisation is a rough approximation. Due to the nonlinear characteristic, there is a threshold expressed by (3.48) and (3.49). Input signals below that threshold are not transferred to the output. Again, the problem is less severe for higher order modulators.

3.5. A COMPARISON OF SIGMA-DELTA MODULATION WITH OTHER DATA CONVERTER TYPES: WHEN TO USE WHAT?

Depending on the factor that limits the Integral Nonlinearity, data converters can be divided in two classes: data converters limited by **component matching** and converters based on **counting algorithms**.

Among the A-to-D converters of the first class, the **flash** ADC of Fig. 3.33.a. [1-5] is the fastest. For a B bit ADC, the input signal is compared with 2^B reference voltages obtained with for instance a resistor string. The digital output word is obtained from the comparator outputs by digital logic. The sampling rate is determined mainly by the comparator settling time and the accuracy is limited by the resistor matching and by the comparator offset voltages. A-to-D converters based on this principle with 8 bit resolution at 100 Msample/sec in a CMOS technology have been reported [3].

Because the number of comparators increases with 2^B, this principle becomes awkward for word lengths over 6 bits. Larger word lengths can be obtained with a **two-step flash** ADC [6-13] as depicted in Fig. 3.33.b. The first flash ADC determines the B_1 most significant bits and the second one converts the residue to B_2 least significant bits. The number of comparators is now $2^{B1} + 2^{B2}$. With pipelining, the speed penalty can be small. The accuracy is limited by component matching, by comparator offset voltages and by the accuracy of the multiplication and the subtraction.

This principle can be extended towards a **pipelined** ADC [14-16] with B one-bit sections as depicted in Fig. 3.33.c. Although the output is delayed over B clock cycles, the throughput equals one output per clock cycle. Here, the accuracy is limited by the multiplication factor which has to be as close to two as possible. A resolution of 12 bit at 1 Msample/sec. based on this principle has been reported [13].

The **recursive** ADC of Fig. 3.33.d. [17-22] re-uses one comparator to determine consecutive bits. Since it takes B clock cycli to produce one B-bit output word, the sampling rate is B times smaller than the internal clock rate. The circuit complexity is reduced at the cost of speed penalty. A 12 bit - 200 ksample/sec. performance has been achieved [19].

A similar technique is applied in the **successive approximation** ADC of Fig. 3.34.e [23]: each clockcycle, an additional bit is determined by comparing the input signal with the DAC output. The performance of this technique is comparable with that of recursive data converters.

For all these converter types, the accuracy is ultimately limited by component matching. In a standard CMOS process, capacitor or resistor matching without trimming yields about 10 to 12 bit.

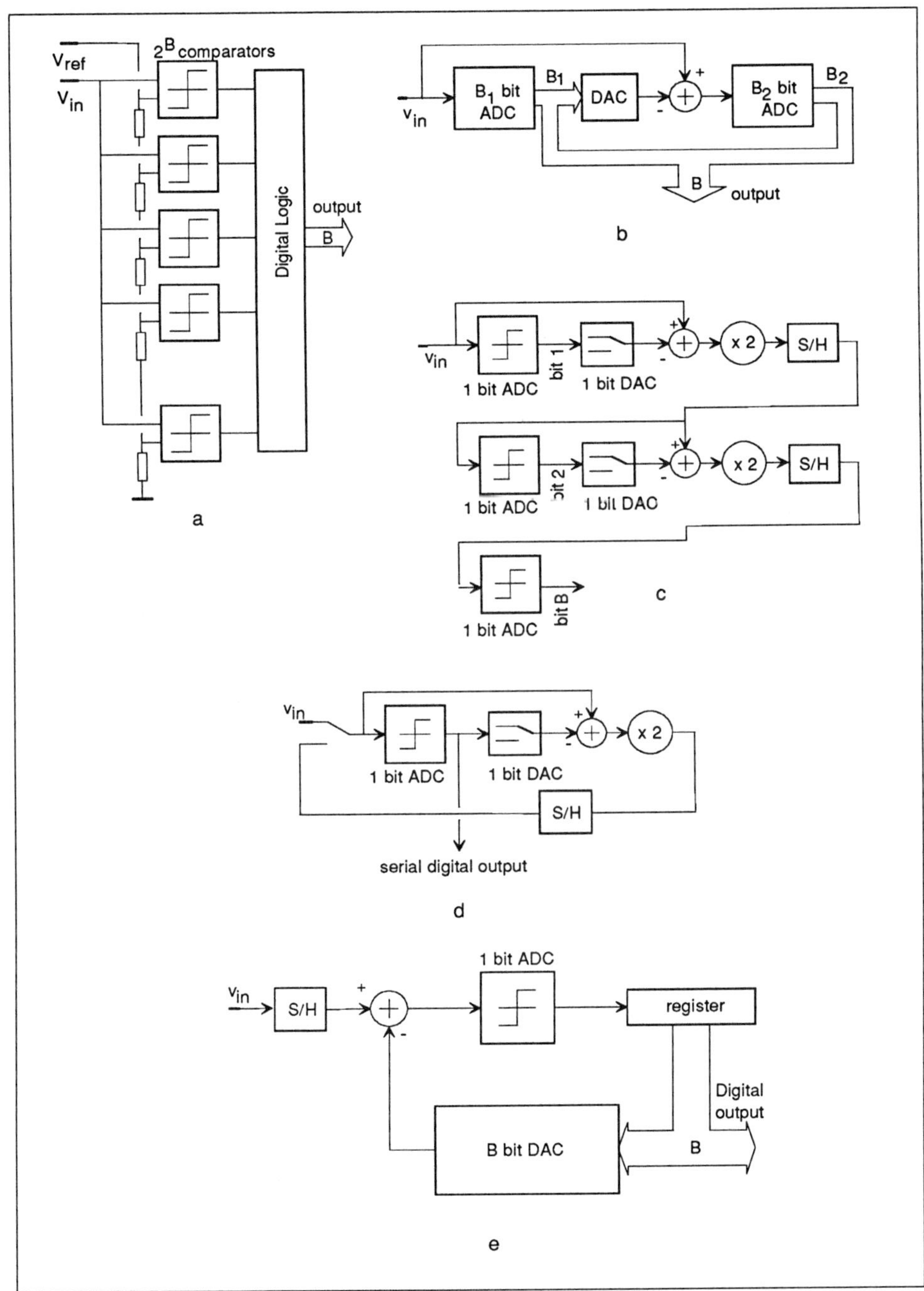

Fig. 3.33: The basic ADC topologies based on component matching

A solution to overcome this limit is to use an ADC based on **counting**, for instance a **dual slope** ADC [47]: a charge, proportional to the analog input is placed on a capacitor. Then the capacitor is discharged with reference charge packages. The number of packages required to discharge the capacitor is a measure for the analog input. This technique can be very accurate but the conversion rate is slow. For instance, to obtain 14 bit accuracy, the worst case number of reference charge packages is 2^{14}. This implies that the conversion rate is 2^{14} times smaller than the internal clock frequency. The dual-slope technique is mainly applied for measurement purposes such as a digital voltmeter where only a few conversions per second are required.

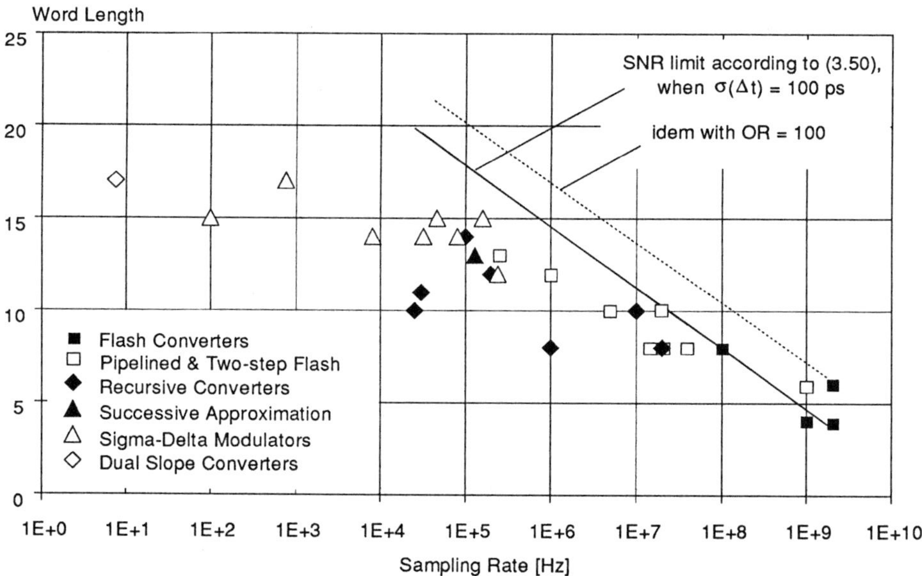

Fig. 3.34: A comparative overview of the different ADC principles

A **Sigma-Delta modulator** [48-53] is also based on a counting algorithm. For this ADC type, the output is determined by counting the number of ones in the PDM signal. When the successive input samples are uncorrelated (this is the case, for instance for a digital voltmeter), the Sigma-Delta modulator offers no advantage over a dual slope ADC. But when the analog input is a sampled band-limited signal (for instance a sampled audio signal), a major part of the quantisation noise power is not in the signal band. For a given accuracy, the internal clock frequency of a Sigma-Delta ADC can be much smaller than that of a dual slope ADC. E.g., while a dual slope ADC requires an internal clock frequency of 2^{14} times the sampling rate to obtain 14 bit accuray, an oversampling ratio of 68 is sufficient for a Second-order Sigma-Delta ADC. Because of its efficient counting algorithm, the trade-off between speed and accuracy is more advantageous for a Sigma-Delta modulator. The counting, which is nothing else than a lowpass function, is performed by the decimator.

D-to-A converters can be classified in a similar way: converters based on component matching, such as binary-weighted current sources or R-2R resistor arrays, are fast but the INL is limited to about 12 bit. D-to-A converters based on counting, such as an algorithmic DAC or a Sigma-Delta DAC, are accurate but slow.

Fig. 3.34 depicts an overview of the A-to-D converters published over the last two years. Flash converters are the appropriate choice for fast converters with word lengths up to 6, at most 8, bits. For a word length up to 10 to 12 bit, pipelined converters are applied for fast applications while recursive converters offer a more economical solution for slower sampling rates. Resolutions over 14 bit without trimming procedures can only be achieved with converters based on counting. For DSP applications, a Sigma-Delta modulator is the best choice.

Besides the efficient counting algorithm, a Sigma-Delta modulator offers some additional advantages:

- As depicted in Fig. 3.23.a, the input signal of an ADC passes through an analog anti-aliasing filter first. Because of the finite filter roll-off, the signal band is smaller than half of the sampling rate (see Fig. 3.35.a.). In a Sigma-Delta modulator, the signal is sampled at a much higher rate than the Nyquist frequency. Only signals above $f_s\text{-}f_d/2$ can be aliased in the signal band. As a result, a smoother passband slope can be tolerated (see Fig. 3.35.b.) which results in a simpler filter design. Spurious signals at frequencies between $f_d/2$ and $f_s\text{-}f_d/2$ are removed afterwards by the decimating filter. Hence, the complex analog filter is replaced by digital circuitry.

- Due to clock jitter, the ADC performance is ultimately limited by aperture noise, expressed by (3.4). With a worst-case signal frequency equal to half of the sampling rate, this yields

$$\text{SNR} = -20.\log[\pi f_s.\sigma(\Delta t)] \text{ dB} \qquad (3.50)$$

This relationship between sampling rate and resolution is depicted in Fig. 3.34 (full line) for a clock jitter of 10 psec. For a Sigma-Delta modulator, only a little part of this aperture noise is in the signal band and (3.50) becomes

$$\text{SNR} = -20.\log[\pi f_d.\sigma(\Delta t)/\sqrt{OR}] \quad \text{dB} \tag{3.51}$$

This relationship is plotted in Fig. 3.34 (dashed line) for an Oversampling ratio of 100. Clearly, a Sigma-Delta modulator is less sensitive to clock jitter.

-   As mentioned in section 3.2.c, the hold operation in a DAC causes a suppression of high-frequency signals. In a classical DAC, the signal band contains frequencies up to $f_s/2$ and |H(f)| in expression (3.9) can be as low as 0.63. For a Sigma-Delta DAC, the signal band is much smaller than half of the sampling frequency and |H(f)| is closer to unity. For instance for an oversampling ratio of 100, |H(f)| varies between 1 and 0.99995.

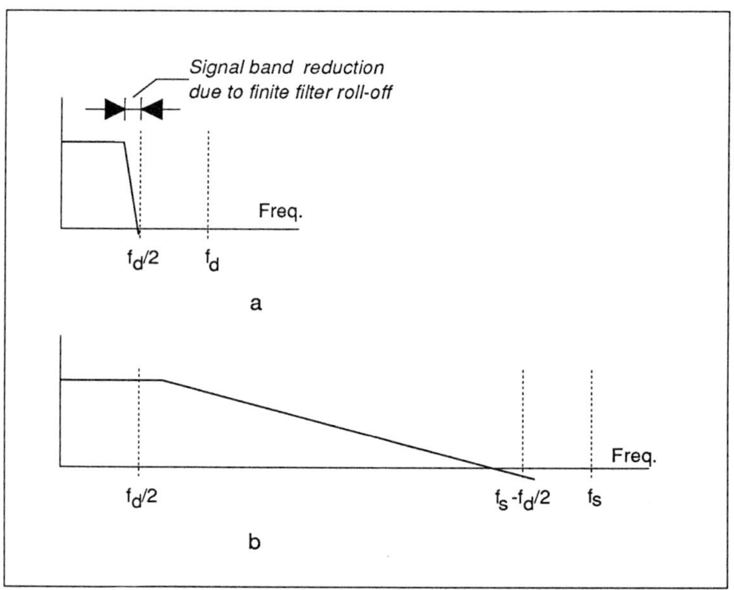

Fig. 3.35: The anti-aliasing filter characteristic for a) a classical ADC and b) an oversampled ADC

Recent self-calibration techniques [54] or dynamic current mirrors [20] allow to realise accurate binary weighted current sources. In principle, this offers a second option to overcome the limit of 10 to 12 bit, imposed by lack of component matching. However, the Sigma-Delta technique still allows an economical solution for sampling frequencies up til a few 100 kHz, just because of the additional advantages mentioned above [30].

3.6. SIGNAL PROCESSING OPERATIONS IN THE PDM DOMAIN

In a Sigma-Delta A-to-D converter as depicted in Fig. 3.23.a, the modulator one-bit PDM output signal is converted into a multi-bit digital signal representation by a decimator. This decimator consists of a digital lowpass filter and a downsampler. Since this book is focusing on the analog section of oversampled data converters, an in-depth discussion of digital decimators is beyond the scope of this work. An introduction can be found in reference [35].

However, besides decimation, several interesting linear data processing operations can be performed on a one-bit PDM signal. For instance, with a very simple digital circuit, two one-bit PDM signals can be summed, resulting in a new one-bit PDM signal. This new PDM signal equals the sum of the two input PDM signals, plus high-frequency quantisation noise. It is also possible to scale a PDM-signal with a fixed factor or to perform a digital filtering on a PDM signal. As will be shown in this section, each of these operations can be performed in the one-bit PDM domain, with relatively simple digital circuits.

3.6.a. Sigma-Delta modulators for unsigned signals

The Sigma-Delta modulators presented in this book up till now are intended for the processing of signed, AC signals which can be positive or negative with respect to an analog ground voltage. For instance, for the A-to-D converter of Fig. 3.36.a, the input signal can vary between $+V_{ref}$ and $-V_{ref}$ and the PDM output signal oscillates between plus one and minus one.

An alternative circuit is shown in Fig. 3.36.b. In this circuit, the input dynamic range spans from $+V_{ref}$ to ground. The PDM output signal oscillates between one and zero, and the DAC output signal is switched between $+V_{ref}$ and ground. This Sigma-Delta modulator is suited for the conversion of unsigned, positive signals. It is easy to show that the previously derived relation (3.27) between SNR, oversampling ratio and modulator order remains valid for this modulator type.

Although all the principles derived further in this section are valid for both signed and unsigned signals, all the signals will be assumed unsigned: analog signals can vary between $+V_{ref}$ and ground, and PDM signals vary between plus one and zero. This assumption will simplify the notations.

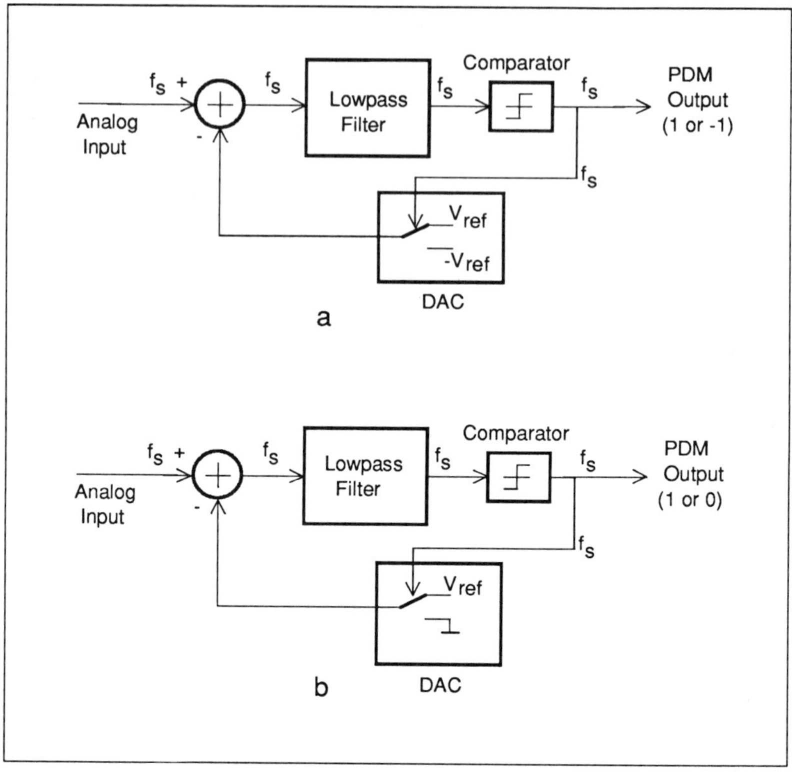

*Fig. 3.36: A Sigma-Delta modulator for a) signed signals
and b) unsigned positive signals*

3.6.b. Scaling the low-frequency content of a PDM signal.

With a relative simple digital circuit, it is possible to scale down the in-band signal content of a one-bit PDM signal. This can for instance be used as an electronic volume control, e.g. in a one-bit CD player: the CD player output signal amplitude can be varied by scaling the PDM-signal with a variable scaling factor.

The way to perform this scaling can best be illustrated with the circuit of Fig. 3.37.a. This circuit consists of a B-bit full adder and a B-bit latch, configured as a digital accumulator. A B-bit digital input signal X(t) is applied to the circuit, and the output is a one bit signal y(t), formed by the B+1-th bit of the sum (the carry-out of the adder).

The signal flow of this circuit is shown in Fig. 3.37.b: when the adder output is smaller than 2^B, i.e. when the carry-out is zero, the full adder output is applied to the latch and fed back to the accumulator during the following clock cycle. On the other hand, when the adder output is not less than 2^B, the adder output minus 2^B is fed back. This operation can be represented by a comparator, as shown in Fig. 3.37.b.

By moving the one-clock delay element (z^{-1} in Fig. 3.37.b.) to the forward path, the equivalent schematic of Fig. 3.37.c. is obtained. When noticing that the circuits inside the dashed rectangle form a digital integrator with transfer function $1/(z-1)$, it becomes clear that this circuit is a First-order Sigma-Delta modulator, as shown in Fig. 3.23.b.

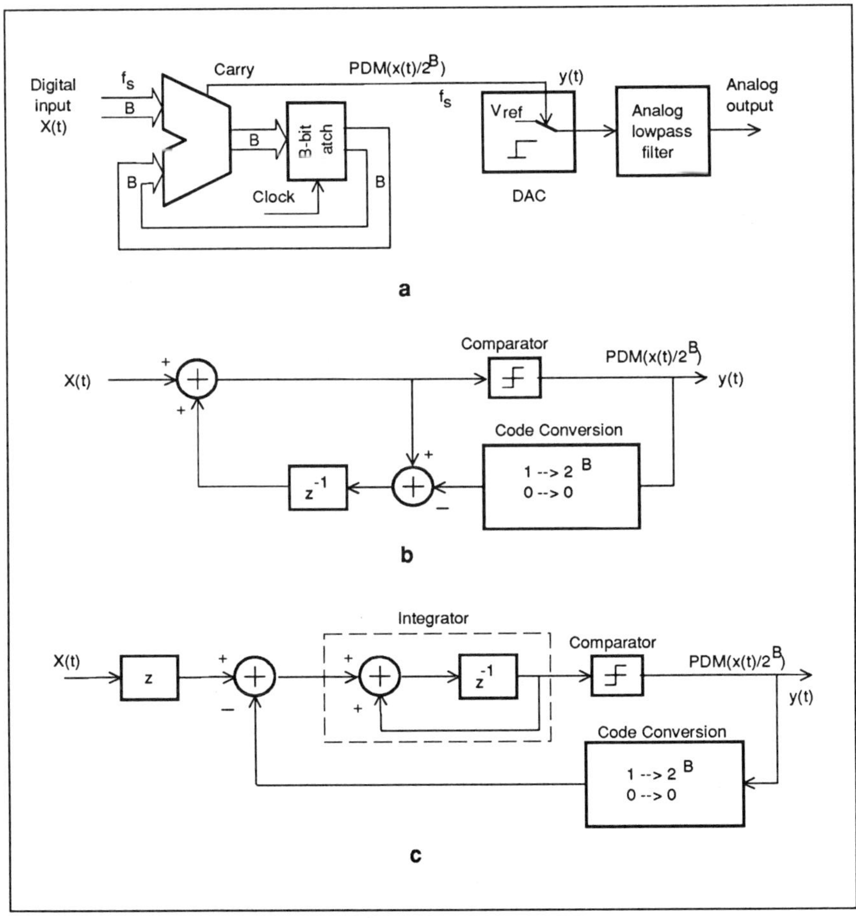

Fig. 3.37: a) a digital Sigma-Delta modulator
b) and c): principle of operation

Hence, the circuit of Fig. 3.37.a. is a First-order digital Sigma-Delta modulator and the output signal y(t) is nothing else than the one-bit PDM representation of the input signal X(t). More accurately, since X(t) ranges from zero to 2^B-1:

$$y(t) \text{ is the PDM representation of } \frac{X(t)}{2^B} \tag{3.52}$$

This implies that y(t) equals the input signal $X(t)/2^B$, plus quantisation noise. This quantisation noise has the noise shaping of a First-order Sigma-Delta modulator.

The same principle can be applied to a one-bit input signal (B equals one), as shown in Fig. 3.38.a. This circuit is a one-bit version of Fig. 3.37.a. The output signal $y_2(t)$ is a PDM representation of the input signal $y_1(t)$ divided by two. When now the input signal $y_1(t)$ itself is the PDM representation of an analog signal x(t), $y_2(t)$ will contain the signal $x(t)/2$, plus high-frequency quantisation noise. In other words, the low-frequency signal content - not the noise - of $y_1(t)$ is scaled with a factor 1/2.

A scaling with other factors can be performed with the circuit of Fig. 3.38.b. Here, for example, the PDM input signal $y_1(t)$ is multiplied first with a factor of five. The output signal $y_2(t)$ equals:

$$y_2(t) = \frac{5 \cdot y_1(t)}{2^3} \text{ plus quantisation noise} \tag{3.53.a}$$

$$= \frac{5}{8} \cdot y_1(t) \text{ plus quantisation noise} \tag{3.53.b}$$

Other scaling factors can be obtained by changing the connections of $y_1(t)$ to the adder input.

With the circuit of Fig. 3.38.c, the scaling factor is programmable:

$$y_2(t) = \frac{Y}{2^B} \cdot y_1(t) \text{ plus quantisation noise} \tag{3.54}$$

where Y is a constant, B-bit digital word. The quantisation noise in expressions (3.53) and (3.54) is shaped as the noise of a First-order Sigma-Delta modulator.

Note that the circuit of Fig. 3.38.c. is much less complicated than a conventional B-by-B-bit multiplier.

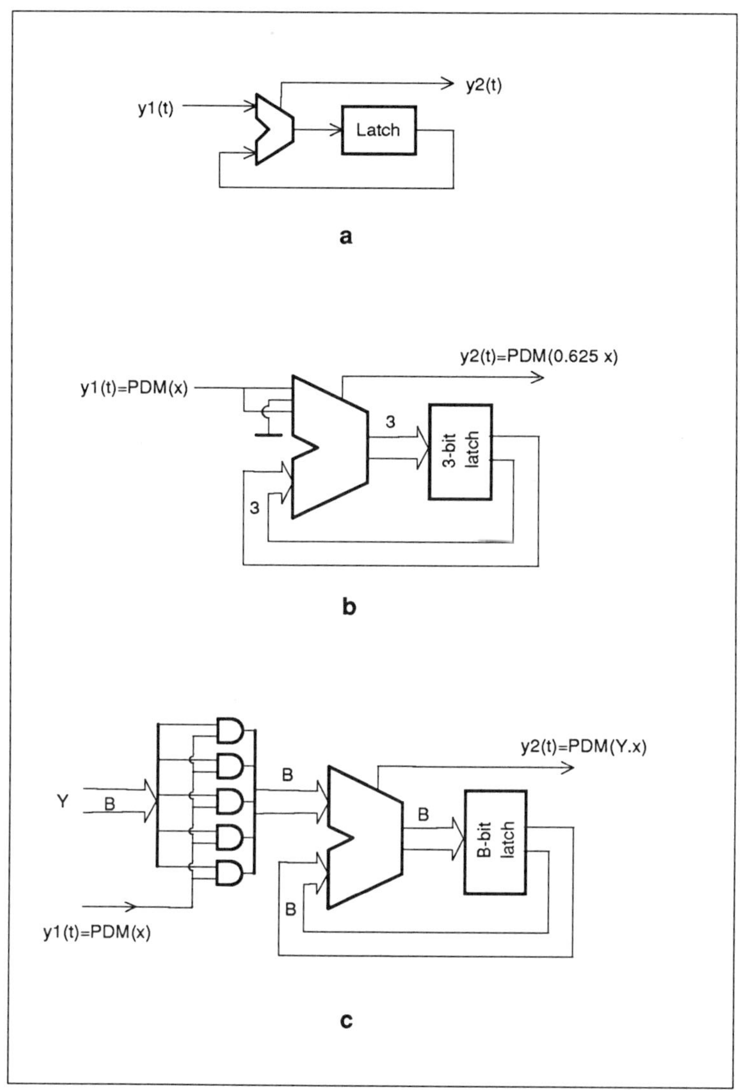

Fig. 3.38: Scaling a PDM signal a) by a factor 1/2, b) by a factor 5/8 and c) by a programmable factor Y.

3.6.c. Summing the low-frequency content of two PDM signals

The circuit of Fig. 3.38.a. can be extended towards the schematic depicted in Fig. 3.39. [71-72] Here, a full-adder is used to allow two one-bit PDM input signals. With a signal flow diagram, similar to Figs. 3.36.b. and 3.36.c, it can be understood that the one-bit output signal $y_3(t)$ is given by:

$$y_3(t) \;=\; \frac{y_1(t) \;+\; y_2(t)}{2} \quad \text{plus quantisation noise} \qquad (3.55)$$

When $y_1(t)$ and $y_2(t)$ are the PDM representations of two analog signals $x_1(t)$ and $x_2(t)$ respectively, the PDM signal $y_3(t)$ contains the signals $x_1(t)/2$ and $x_2(t)/2$, plus quantisation noise.

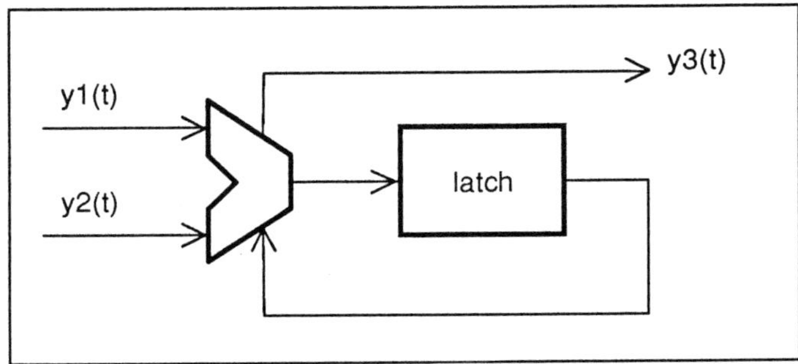

Fig. 3.39: A circuit to add two PDM signals.

3.6.d. Filtering the low-frequency content of a PDM signal

Digital filters consist of delay elements, adders and signal scaling elements. Since all these operations can be performed in the one-bit PDM domain, it is possible to filter the in-band signal content of a PDM signal [72]. For instance, the circuit of Fig. 3.40. is a lowpass comb filter with transfer function:

$$H(z) \;=\; \frac{1}{16} \;+\; \frac{1}{8}.z^{-1} \;+\; \frac{1}{4}.z^{-2} \;+\; \frac{1}{2}.z^{-3} \qquad (3.56)$$

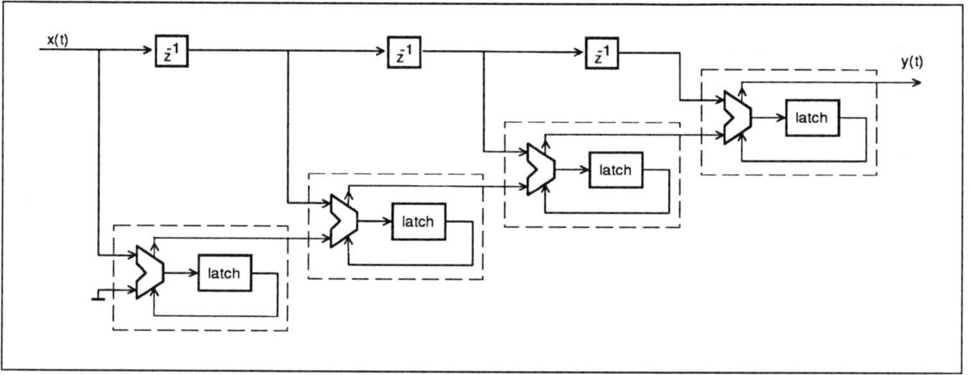

Fig. 3.40: a one-bit PDM lowpass filter

3.7. SIMULATING SIGMA-DELTA MODULATORS

Calculations of the in-band noise (see expression (3.25)) or the system poles (see (3.30)) relying on the white noise assumption or the linearised gain k_1 yield poor agreement with the reality. Therefore, computer simulations are the only reliable way to determine the output power spectrum and the SNR of a Sigma-Delta loop. Since the signal band is much smaller than the clock frequency, a large number of clock periods have to be calculated to obtain one signal period. With classical circuit simulators such as SPICE, this would require unrealistic computer efforts [*].

The n-th order Sigma-Delta modulator loop of Fig. 3.21. can be simulated with a behavioural simulator such as SABER [65] or with the following dedicated routine (programmed in C or PASCAL):

```
for k = 1 to order          /* set to initial conditions */
   v(k) = IC(k)
end
for time = 1 to simulationlength    /* simulate */
   x = dcbias + amplitude*sin(2*pi*time/period)
   if v(order) > 0 then y(time)=1 else y(time)=-1
     for k = order to 2 step -1
       v(k) = v(k) + a(k)*(v(k-1)-b(k)*y(time))
     end
   v(1)=v(1) + a(1)*(x-y(time))
end
"perform fft on vector y"
"Calculate in-band noise power"
"Calculate SNR"
```

(*) This situation is similar to the simulation of Switched Capacitor filters. There, the problem is solved with dedicated high-level simulators such as SWAP [55] or SWITCAP [56].

where "x" is the input voltage consisting of a sinus superimposed upon a DC bias. "y" is the comparator output and "v(k)" is the k-th integrator output voltage. In the first "for-loop", the integrator output voltages are set to their initial conditions "IC(k)". The analog voltages are scaled to V_{ref} in order to obtain a number with no dimension. "order" is the order of the Sigma-Delta modulator, "a_k" is the k-th integrator gain and b_k is the k-th feedback coefficient.

In a similar way, the influence of non-ideal building blocks can be investigated. For instance, leaky integrators can be modelled by replacing the eighth line of the routine by

```
v(k) = [v(k) + a(k)*(v(k-1)-b(k)*y(time))] /[1+a(k)*leak(k)]
```

where "leak(k)" is the leakage of the k-th integrator.

In order to avoid the generation of a new simulation program for each convertor type, dedicated high-level behavioural simulators for oversampled data converters such as MIDAS [59] or TOSCA [62] are available or under development.

Fig. 3.41 shows the simulated PDM output spectrum of a First-order Sigma-Delta modulator driven with a large sinusoidal signal. The upper line presents the total in-band noise power, i.e. when for instance the PDM signal is decimated with a lowpass filter frequency of $0.01.f_s$, the in-band noise power is -58 dB.
The input signal is a large sinusoidal signal is this simulation. At each zero-crossing of the modulator input signal, the in-band noise power increases due to pattern noise (see Fig. 3.30.a). As a result, this spectrum shows large quantisation noise peaks at harmonics of the input signal frequency. Note however, these peaks are below the noise floor.

In Fig. 3.42, similar curves are plotted for the Second-order Sigma-Delta modulator. As expected, the in-band noise power (see the upper line) is smaller for filter bandwidths below 0.05 times the sampling frequency (i.e. OR larger than 10). The low-frequency noise peaks are about 20 dB smaller than those of the first-order modulator.

Fig. 3.43. shows the output spectrum of a Fourth-order modulator. As can be seen, the in-band noise power is further reduced for oversampling ratios higher than 40. The spectrum contains no peaks at harmonics of the input signal frequency. This illustrates again that increasing the modulator order reduces the pattern noise.

In Fig. 3.44, the SNR obtained from the simulations of Fig. 3.41 and 3.37 is compared with the predictions of expression (3.27). Despite the unreliable white noise assumption and the quantiser linearisation, the calculations agree remarkably well with the simulation for a First-order modulator. This explains why the white noise assumption is so popular. For the Second-order modulator, expression (3.27) yields the correct slope but the predicted values are about 10 dB too optimistic.

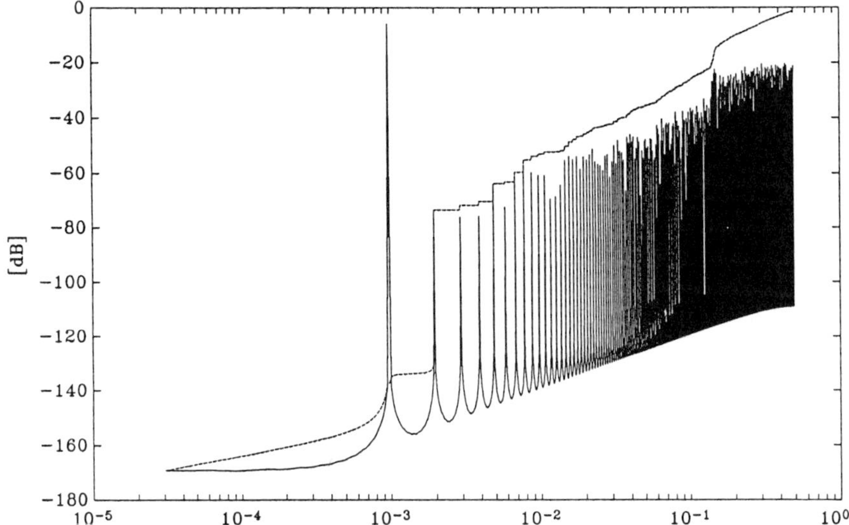

Fig. 3.41: The simulated output power spectrum of a First-order Sigma-Delta modulator driven by a large sinusoidal input signal.
output spectrum (lower line) and in-band noise power (upper line). The largest peak is the signal, the others are pattern noise

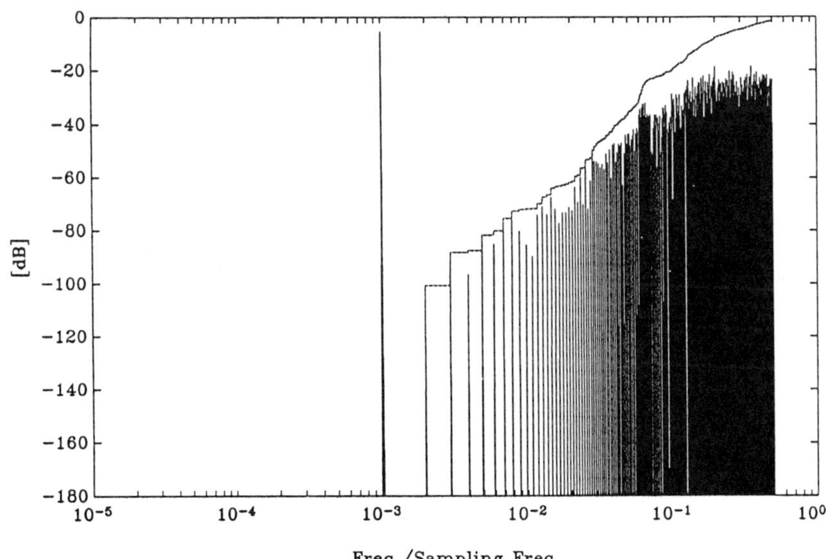

Fig. 3.42: The simulated output power spectrum of a Second-order Sigma-Delta modulator. output spectrum (lower line) and in-band noise (upper curve)
The largest peak is the signal, the others are noise

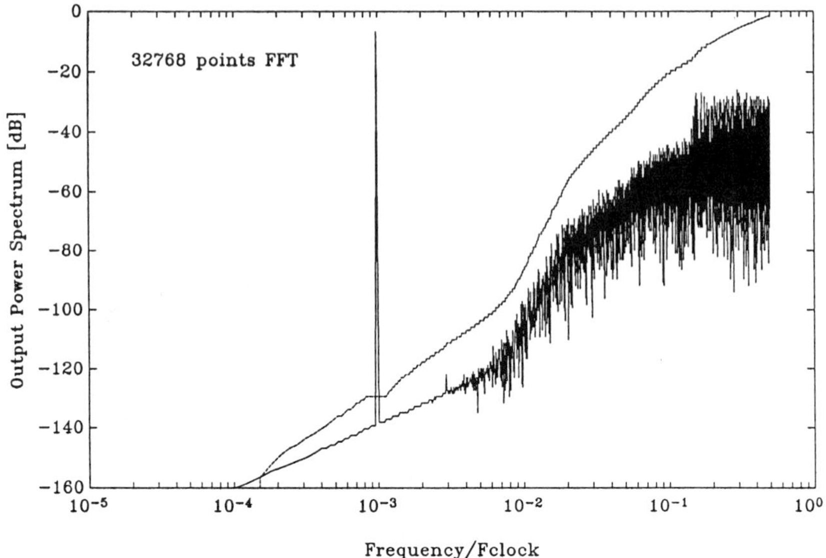

Fig. 3.43: The output spectrum of a Fourth-order modulator driven by a large sinusoidal signal.
quantisation noise spectrum (lower line) and in-band noise power (upper curve)

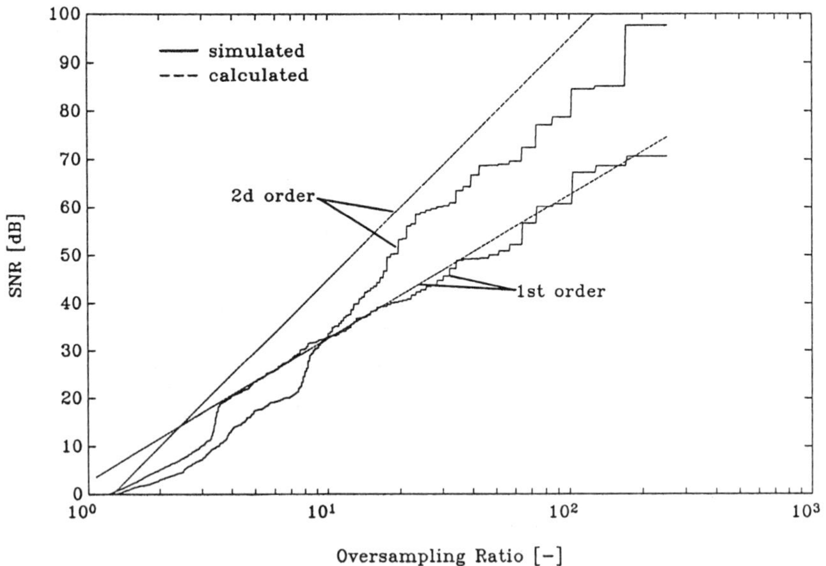

Fig. 3.44: Simulated versus Calculated SNR characteristics for a First- and Second-order Sigma-Delta modulator

3.8. TESTING A/D AND D/A DATA CONVERTERS

Data converters intended for static operation (e.g. the ADC in a digital voltmeter) can be characterised with a nonlinear DC transfer characteristic as depicted in Fig. 3.10. The Integral and Differential Nonlinearity [57-58] are important benchmarks for this application.

The quality of data converters intended for signal processing applications on the other hand is determined by the SNR and the nonlinear distortion of dynamic signals, rather than by static characteristics. Hence, these data converters are normally tested with dynamic input signals.

A classical setup to determine the output signal quality of an ADC is depicted in Fig. 3.45.a. The ADC-under-test is driven by a sinusoidal input voltage from a reference oscillator with a frequency on the edge of the signal band. The digital output is re-converted to analog. After filtering out the fundamental frequency with for instance a HP339 distortion analyser, the analog output signal is applied to a Signal analyser such as a HP3562A where the quantisation noise and the distortion components can be visualised. The bottle neck in this setup is the SNR of the DAC, which should be at least an order of magnitude higher than the SNR of the ADC-under-test.

A better approach is depicted in Fig. 3.45.b. where the digital output signal is stored in a RAM buffer. After the measurement, the buffer content is transferred to a host mainframe computer where the digital signal is postprocessed [59]. With a Fast Fourrier Transform, the quantisation noise and the distortion components are analysed. From the power at frequencies other than the main signal frequency, the total noise power and the SNR can be calculated.

With a similar approach, the performance of a DAC can be tested as illustrated in Fig. 3.45.c: A digital input pattern is generated on a host computer, stored in a RAM buffer and applied to the DAC input. The analog output is fed to a distortion analyser (e.g. a HP339) where the fundamental frequency is filtered out. The remaining signal containing the quantisation noise and the nonlinear distortion signals can be visualised with an analog signal analyser or with an oscilloscope.

In general, a breadboard hardware setup is required to perform the digital data aquisition and storage (the block 'Buffer RAM' in Fig. 3.45).

Since there is no synchronisation between the clock generator and the signal generator of Fig. 3.45.b, the buffer content sent to the host computer is a signal sample of finite time extent and it normally contains not exactly an integer number of signal periodes as demonstrated in Fig. 3.46.a. It is well known [63-64] that this deteriorates the results of the Fast Fourier Transform.

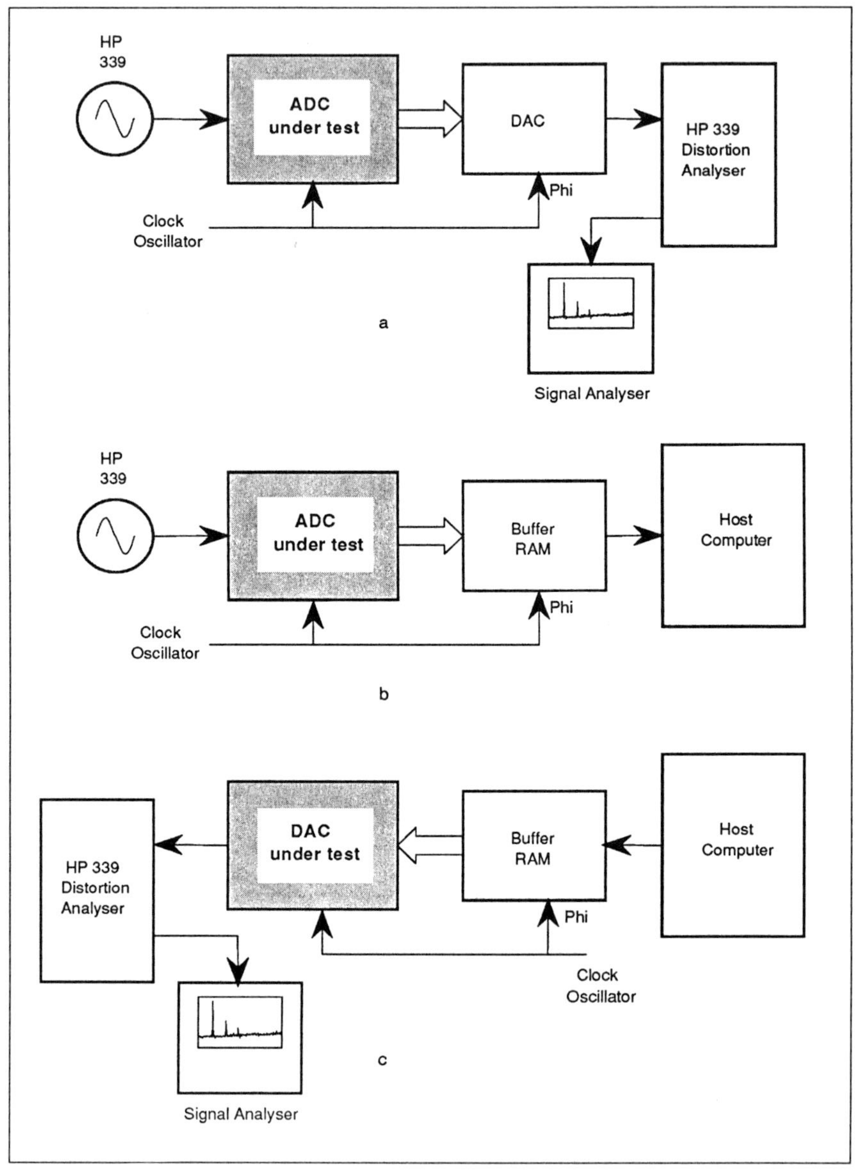

Fig. 3.45: Different data converter measurement setups

In order to illustrate this, Fig. 3.46.b. shows the FFT power spectrum of a sine wave when the buffer contains 131.072 signal periods. Instead of one line, the FFT consists of a wide spectrum. This is **spectral leakage.** When small signals (for instance quantisation noise) are present at other frequencies than the signal frequency, their spectrum is masked by the leakage from the main signal and it is impossible to retrieve the SNR. Hence, an FFT obtained from the finite time sample is not a good estimate for the power spectrum of the original signal.

This problem can be solved by using a **window** [27]: before calculating the FFT, the samples in the buffer are multiplied with a weight function w(n). For instance, the "Hanning" window (or "Raised Cosine" window) is very popular and available in almost every signal analyser. This window multiplies the n-th signal sample with

$$w(n) = 1 - \cos(2\pi n/N) \tag{3.57}$$

where N is the total number of samples in the buffer. The resulting power spectrum is shown in Fig. 3.46.c. With the Hanning window, the spectral leakage components are at least -32 dB below the main signal frequency component. Therefore, this window is sufficient for the testing of A-to-D converters with a word length up to 6 bit.

The "4-term Blackmann-Harris window" is useful for data converters with a higher word length. The weight function of this window is given by

$$w(n) = 0.35875.\cos(2\pi n/N) - 0.48829.\cos(4\pi n/N) \\ + 0.14128.\cos(6\pi n/N) - 0.01168.\cos(8\pi n/N) \tag{3.58}$$

and the largest leakage component is below -92 dB. Fig. 3.46.d. depicts the power spectrum of the buffer content shown in Fig. 3.46.a, weighted with the 4-term Blackmann-Harris window. As can be seen, the leakage components are much smaller than in Fig. 3.46.b. and c.

Selecting the appropriate window for a specific application from the large number of windows described in literature is not a straightforward task. Also, the level of the leakage components is not the only criterion: by applying a window, the SNR of the signal is deteriorated and the spectral resolution is decreased. A detailed discussion of all these effects is beyond the scope of this text. An excellent overview can be found in [63] by F.J. Harris. Note however that the application of the 4-term Blackmann-Harris window deteriorates the SNR with 3.8 dB or less. This window yields a fair compromise between spectral leakage and SNR degradation. It is a good general-purpose choice for high-performance data converter measurements.

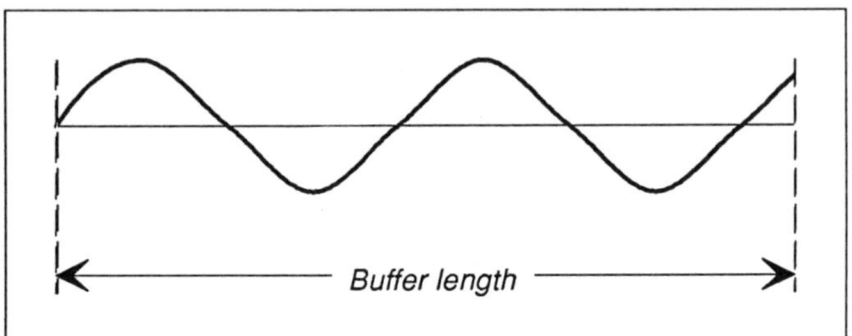

Fig. 3.46.a: Buffer content is not an integer number of signal periods

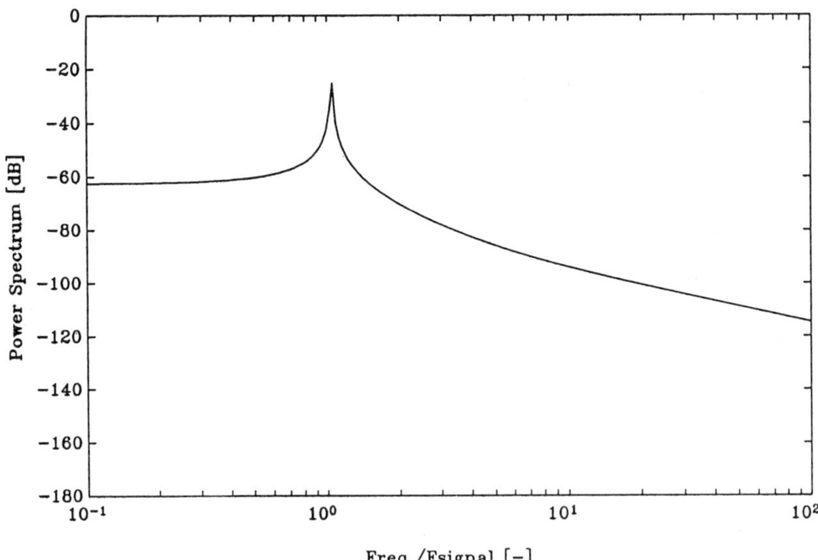

Fig. 3.46.b: The power spectrum of a sine with a flat window

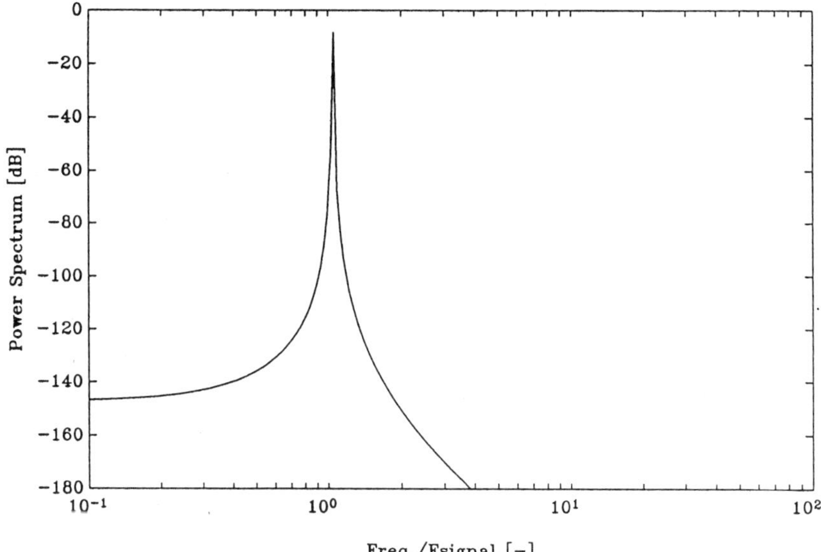

*Fig. 3.46.c: The power spectrum of a sine, estimated with
the Hanning window*

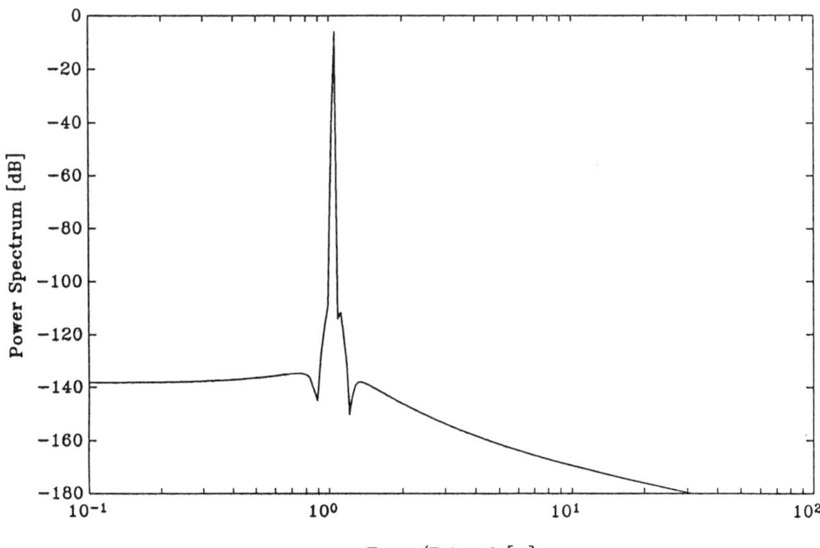

*Fig. 3.46.d: The power spectrum of a sine, estimated with
the 4-term Blackmann-Harris window*

3.9. SUMMARY

For word lengths larger than about 12 bit, oversampled data converters are an alternative for classical data converters relying on component matching. Oversampling yields a high accuracy at the cost of a speed reduction. Moreover, the oversampling technique allows to relax the requirements for the anti-aliasing filter, the smoothing filter and the clock jitter, at the cost of an increased digital circuit complexity.

Because of the nonlinear quantiser transfer characteristic, the SNR of a Sigma-Delta modulator is hard to predict. Simplified calculations relying on the white noise assumption and the linearised gain approximation yield a poor agreement with the reality for modulator orders higher than one. Hence, high level simulations are necessary to predict the modulator characteristics. Moreover, secondary effects such as pattern noise or low signal-level effects have to be taken into account. First-order Sigma-Delta modulators are seldomly used, mainly because of these secondary effects.

While the digital circuits of Fig. 3.23. are more of concern for digital circuit engineers, the analog Sigma-Delta loop of Fig. 3.23.a. and the analog reconstruction filter of Fig. 3.23.b. are of interest for an analog circuit designer. A detailed study of the design requirements for these two building blocks is the subject of the two following chapters.

3.10. REFERENCES

[1]   R. VAN DE PLASSCHE, P. BALTUS: "An 8b Folding ADC" - *IEEE 1988 ISSCC Digest of Techn. papers* pp. 222-223

[2]   T. WAKIMOTO *et al:* "Si Bipolar 2Gs/s 6b Flash A/D Conversion LSI" - *IEEE 1988 ISSCC Digest of Techn. papers* pp. 232-233

[3]   R. J. VAN DE PLASSCHE *et al:* "An 8-bit 100 MHz Full Nyquist Analog-to-Digital Converter" - *IEEE J. Solid-State Circuits* vol. SC-23 no.6, Dec. 1988 pp. 1334-1339

[4]   V. E. GARUTS *et al:* "A Dual 4-bit 2 Gs/s Full Nyquist Analog-to-Digital Converter Using a 70-ps Silicon bipolar Technology with Borosenic-Poly Process and Coupling-Base Implant" - *IEEE J. Solid-State Circuits* vol. SC-24 no.2, april 1989 pp. 216-222

[5]   D. DANIEL *et al:* "A Silicon Bipolar 4-bit 1 Gsample/sec. Full Nyquist A/D Converter" - *IEEE J. Solid-State Circuits* vol. SC-23 no.3, June 1989 pp. 742-749

[6]   J. DOERENBERG, D. A. HODGES: "A 10-bit 5Msample/sec CMOS 2-Step Flash ADC" - *Proc. IEEE CICC 1988* pp. 18.6.1-4

[7]   N. FUKUSHIMA *et al:* "A CMOS 40 MHz 8b 105 mW Two-Step ADC" - *IEEE 1989 ISSCC Digest of Techn. papers* pp. 14-15

[8]   T. MATSURA *et al:* "An 8b 20 MHz CMOS Half-Flash A/D Converter" - *IEEE 1988 ISSCC Digest of Techn. papers* pp. 220-221

[9]   A. CREMONESCI *et al:* "An 8-bit Two-Step Flash A/D Converter for Video Applications" - *Proc. IEEE CICC 1989* pp. 6.3.1-4.

[10]   J. DOERNBERG, P. R. GRAY, D. A. HODGES: "A 10-bit 5-Msample/sec. CMOS Two-step Flash ADC" - *IEEE J. Solid-State Circuits* vol. SC-24 no.2, april 1989 pp. 241-249

[11]   D. A. KERTH *et al:* "A 12-bit 1Msample/sec. Two-Step Flash ADC" - *IEEE J. Solid-State Circuits* vol. SC-24 no.2, april 1989 pp. 250-255

[12]   T. SHIMIZU *et al:* "A 10b 20 MHz Two-step Parallel ADC with internal S/H" - *IEEE 1988 ISSCC Digest of Techn. papers* pp. 224-225

[13]   BANG-SUP SONG, M. F. TOMPSETT: "A 12b 1 MHz Capacitor Error Averaging Pipelined A/D Converter" - *IEEE 1988 ISSCC Digest of Techn. papers* pp. 226-227

[14]   S. SUTARJA *et al:* "A 250 ks/s 13b Pipelined A/D Converter" - *IEEE 1988 ISSCC Digest of Techn. papers* pp. 228-229

[15]   J.-G. CHERN, A. A. ABIDI: "An 11 bit, 50 kSample/s CMOS A/D Converter Cell Using a Multislope Integration technique" - *Proc. IEEE CICC 1989* pp. 6.2.1-4.

[16]   F. THOMAS *et al:* "1-GHz GaAs ADC Building Blocks" - *IEEE J. Solid-State Circuits* vol. SC-24 no.2, april 1989 pp. 223-228

[17]   S. CHIN *et al*: "A Multistep ADC Family with Efficient Architecture" - *IEEE 1989 ISSCC Digest of Techn. papers* pp. 16-17

[18]   H. ONODERA: "A Cyclic A/D Converter that does not Require Ratio-Matched Components" - *IEEE J. Solid-State Circuits* vol. SC-23 no. 1, Febr. 1988 pp. 152-158

[19]   M. YOTSUYANAGI *et al:* "A 12 bit 5µsec CMOS Recursive ADC with 25 mW Power Consumption" - *Proc. IEEE CICC 1989* pp. 6.4.1-4.

[20]   D. G. NAIRN, C. A. T. SALAMA: "Ratio-Independent Current Mode Algorithmic Analog-to-Digital Converters" - *Proc. ISCAS 1989* pp. 250-253

[21]   S. OGAWA *et al:* "A Buffer-Based Algorithmic Analog-to-Digital Converter" - *Proc. ISCAS 1989* pp. 276-279

[22]   J. FERNANDES *et al*: "A 14b 10µs Subranging A/D Converter with S/H" - *IEEE 1988 ISSCC Digest of Techn. papers* pp. 230-231

[23]   S. RAMET: "A 13 bit, 160 kHz Differential Analog to Digital Converter" - *IEEE 1989 ISSCC Digest of Techn. papers* pp. 20-21

[24]   F. DE JAGER: "Delta modulation, a method of PCM transmission using 1-unit code" - *Philips Res. Rep,* vol. 7, Nov. 1952, pp. 442-448

[25]   J. C. CANDY: "A use of Limit Cycle Oscillations to obtain robust analog-to-digital Converters" - *IEEE Trans. Commun.* vol. COM-22 pp.298-305, March 1974

[26]   R. J. VAN DE PLASSCHE: " A Sigma-Delta Modulator as an A/D Converter" - *IEEE Trans. on Circuits and Systems* vol. CAS-25 no.7, pp. 510-514, July 1978

[27]   S.A. TRETTER: "Introduction to Discrete-Time Signal Processing" - Wiley and Sons, New York, 1976

[28]   R. E. CROCHIERE, L. R. RABINER: "Interpolation and Decimation of Digital Signals - A Tutorial Review" - *Proc. of the IEEE* vol. 69, no.3, pp. 300-331, March 1981

[29]   B. BOSER and B. WOOLEY: "The design of Sigma-Delta modulation analog-to-digital Converters" - *IEEE J. Solid-State Circuits* vol. SC-23, Dec 1988, pp. 1298-1308

[30]   E. C. DIJKMANS *et al:* "Sigma-Delta versus Binary weighted AD/DA conversion, what is the most promising?" - *Proc. ESSCIRC* 1989 pp. 35-63

[31]   B. P. AGRAWAL, K. SHENOI: "Design methodology for Sigma-Delta modulators" - *IEEE Trans. Commun.* vol COM-31 pp. 360-370, March 1983

[32]   J. C. CANDY: "A Use of Double Integration in Sigma-Delta Modulation" - *IEEE Trans. Commun.* vol. COM-33 pp.249-258, March 1985

[33]   J. C. CANDY, AN-NI HUYNH: "Double Interpolation for D/A", *IEEE Trans. Commun.* vol. COM-34 pp.77-87. Jan. 1986

[34]   J. C. CANDY: "Decimation for Sigma-Delta modulation" - *IEEE Trans. Commun.* vol. COM-34 pp.72-76, Jan. 1986

[35]   E. DIJKSTRA *et al:* "Wave Digital Decimation Filters in Oversampled A/D Converters" - *Proc. ISCAS 1988* pp. 2327-2330

[36]   E. DIJKSTRA, M. DEGRAUWE *et al*: "A design methodology for decimating filters in Sigma-Delta A/C converters" - *Proc. ISCAS 1987* pp. 479-482

[37]   E. F. STIKVOORT: "Some remarks on the stability and performance of the noise shaper or Sigma-Delta modulator" - *IEEE Trans. Commun.* vol. COM-36 pp. 1157-1162, Oct. 1988

[38]   A. GELBE, W. VANDER VELDE: "Multiple-Input Describing Functions and Nonlinear System Design" - McGraw-Hill, NY, 1968

[39]   S. R. ARDALAN, J. J. PAULOS: "An Analysis of Nonlinear Behaviour in Sigma-Delta Modulators" - *IEEE Trans. Circuits and Systems* vol. CAS-34, pp.593-603, June 1987

[40]   P. VANDELOO, H. VANDENDOOREN: "Studie en Ontwerp van een 12 bit Sigma-Delta modulator" - Thesis K.U.Leuven, 1989

[41]   J. C. CANDY and O. J. BENJAMIN: "The structure of Quantisation noise from Sigma-Delta modulation" - *IEEE Trans. Commun.* vol. COM-29 pp. 1316-1323, Sept. 1981

[42]   V. FRIEDMAN: "The Structure of the Limit Cycles in Sigma-Delta Modulation" - *IEEE Trans. Commun.* vol. SC-36, No.8, Aug. 1988 pp. 972-979

[43]   B. BOSER, B. WOOLEY: "Quantisation error Spectrum of Sigma-Delta modulators" - *Proc. ISCAS* 1988, pp. 2321-2324

[44]   R. GRAY: "Oversampled Sigma-Delta modulation" - *IEEE Trans. Commun.* vol. COM-35, pp. 481-489, May 1987

[45]   G. ZAMES, N.A. SHNEYDOR: "Dither in Nonlinear Systems" - *Trans. on Automatic Control*, vol. AC-21 No.5, October 1976, pp. 660-667

[46]   P. NAUS *et al:* "Low Signal-level Distortion in Sigma-Delta Modulators" - 84th Audio Eng. Soc. Convention, March 1988, Paris

[47]   B. J. RODGERS, C. R. THURBER: "A Monolithic 5.5 digit BiMOS A/D Converter" - *IEEE J. Solid-State Circuits* vol. SC-24 no.3, June 1989 pp. 617-626

[48]   J. C. CANDY, B. A. WOOLEY and O. J. BENJAMIN: "A voiceband Codec with Digital filtering" - *IEEE Trans. Commun.* vol. COM-29, June 1981, pp. 815-830

[49]   S. R. NORSWORTHY, I. G. POST, H. S. FETTERMAN: "A 14-Bit 80 KHz Sigma-Delta Converter" - *IEEE J. Solid-State Circuits* vol. SC-24 pp.256-266, April 1989

[50]   M. REBESCHINI, *et al*: "A high-resolution CMOS Sigma-Delta A/D converter with 320 kHz output rate" - *Proc. IEEE CICC 1989* pp. 6.1.1-4.

[51]   R. KOCH *et al:* "A 12-bit Sigma-Delta analog-to-digital Converter with 15 MHz Clock rate" - *IEEE J. Solid-State Circuits* vol. SC-21 no.6, pp. 1003-1010, Dec. 1986

[52] Y. MATSUYA *et al:* "A 16-bit Oversampling A-to-D Conversion Technology using Triple-Integration Noise Shaping" - *IEEE J. Solid-State Circuits* vol. SC-22 no.6 pp.921-929, Dec. 1987

[53] J. ROBER, P. DEVAL: "A Second-Order High-Resolution Incremental A/D Converter with Offset and Charge Injection Compensation" - *IEEE J. Solid-State Circuits* vol. SC-23 no.3, June 1988 pp. 736-741

[54] W. GROENEVELD, H. SCHOUWENAARS, H. TERMEER: "A Self Calibration Technique for Monolithic High-resolution D/A Converters" - *IEEE 1989 ISSCC Digest of Techn. papers* pp. 22-23

[55] SILVAR-LISCO Company: "SWAP, A Switched Capacitor Network Analysis Program", Heverlee, Belgium, 1983

[56] FANG, Y.TSIVIDIS, Y.P. WING: "A Switched Capacitor Netword Analysis Programme" - Parts I-II, *IEEE Trans. on Circuits and Systems* vol. SC-28, 1983, pp. 4-10 and 41-46

[57] P. O' LEARY, F. MALOBERTI: "Bit Stream Adder for Oversampling Coded data" - *Electronic Letters*, 27th September 1990, vol. 26, No. 20

[58] N. KOUVARAS: "Operations on delta-modulated signal and their application in the realisation of digital filters" - *The Radio and Electronic Engineer*, Institution of Electronic and Radio Engineers, Vol. 48, No. 9, pp. 431-438, September 1978

[59] B. E. BOSER *et al*: "Simulating and Testing Oversampled Analog-to-Digital Converters" - *IEEE Trans. on Computer-Aided Design* vol. 7 no.6 pp. 668-674, June 1988

[60] J. DOERNBERG, H. LEE, D. HODGES: "Full Speed Testing of A/D Converters" - *IEEE J. Solid-State Circuits* vol. SC-19 no.6, dec. 1984 pp. 820-827

[61] M. VANDEN BOSSCHE, J. SCHOUKENS, J. RENNEBOOG: "Dynamic Testing and Diagnostics of A/D Converters" - *IEEE Trans. on Circuits and Systems* vol. CAS-33, no.8, Aug. 1988 pp.775-785

[62] V. F. DIAS, V. LIBERALI, F. MALOBERTI: "TOSCA: a simulator for oversampling converters with behavioural modeling", *Proc. CompEuro '91,* Bologna, Italy, pp. 467-471, May 1991

[63] F. J. HARRIS: "On the Use of Windows for Harmonic Analysis with the Discrete Fourier Transform" - *Proc. of the IEEE* vol. 66, No.1, January 1978, pp. 51-83

[64] Y. C. YENQ: "Measuring harmonic distortion and the noise floor of an A/D converter using spectral averaging" - *IEEE Trans. on Instrumentation and Measurements* vol. IM-37, No. 4, 1988, pp. 525-538

[65] "SABER user's guide" - Analogy Inc, 1989

4

HIGHER-ORDER SIGMA-DELTA A-TO-D CONVERTERS

4.1. INTRODUCTION

The principle of a Sigma-Delta modulator and its application for data converters are discussed in the previous chapter. It is shown there that the Sigma-Delta technique allows to realise precise A-to-D converters, independent of component matching properties. However, the practical implementation of this principle is still complicated by the two following problems:

First, for First- and Second-order Sigma-Delta modulators, the speed-for-accuaracy trade off is not so advantageous and therefore, a high accuracy or a large signal bandwidth results in a huge clock frequency. This requires fast, precise analog circuits which are difficult to design and consume a large amount of power. To overcome this need, modulators with more than two integrators are required. The stability of such a higher-order modulator is only briefly discussed in chapter 3. In this chapter 4, section 2, the stability conditions of higher-order modulators are studied more in detail. It is shown that these higher-order modulators are only conditionally stable. One technique to overcome this stability problem will be discussed in detail.

A second problem is the signal degradation due to analog circuit imperfections. At the discussion of the Sigma-Delta modulator of Fig. 3.21, ideal analog components were assumed. The signal was degraded only by the quantisation noise and by the aperture noise due to clock jitter. In a real design however, there is an additional signal degradation due to circuit imperfections such as thermal noise, component nonlinearities, integrator leakage or clock feedthrough. Especially for those applications with a high dynamic range that require a Sigma-Delta modulator, these imperfections become very important. In section 3, the signal degradation due to the major circuit effects is discussed. Relations are established between the over-all system performance and the design requirements for the individual building blocks. Design techniques are presented to overcome the major signal-degenerating second-order effects.

In section 4, a design example of a Fourth-order Sigma-Delta modulator is presented, incorporating the previously discussed design techniques.

Besides the approach presented in Section 2, several other techniques to ensure the stability of higher-order modulators are described in the literature. An overview is given in Section 5.

4.2. THE STABILITY OF SIGMA-DELTA MODULATORS WITH AN ORDER LARGER THAN TWO

As demonstated in chapter 3, the trade off between the speed penalty and the accuracy of a Sigma-Delta modulator becomes more advantageous with increasing modulator order. Increasing the modulator order also reduces the sensitivity for pattern noise (see section 3.4.a). However, as is shown in this section, the implementation of modulators with an order larger than two is complicated by stability problems.

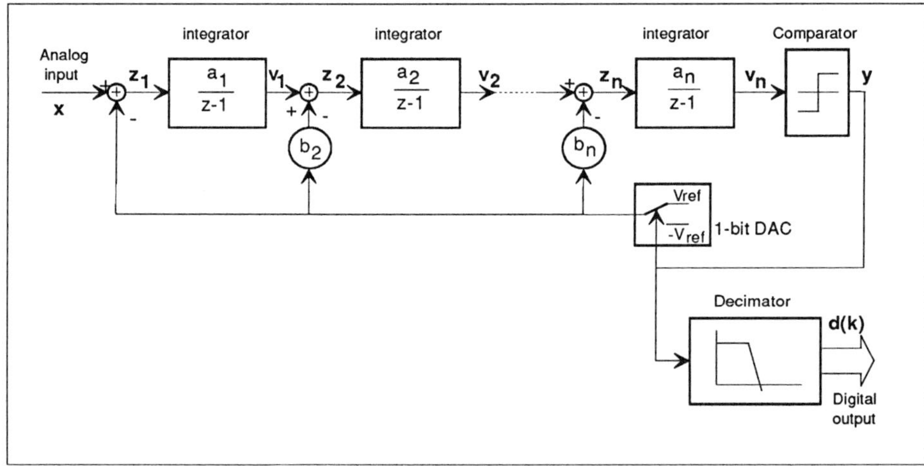

Fig. 4.1: The basic schematic of a Sigma-Delta ADC

The basic schematic of an n-th order Sigma-Delta ADC is repeated in Fig. 4.1. The modulator loop is a nonlinear system with one input signal (x), one output signal (y) and with n state variables (the integrator output voltages v_i). The characteristics of this system are functions of the circuit parameters a_i and b_i.

When a Sigma-Delta modulator is driven with a constant input signal between -Vref and $+V_{ref}$, the PDM output signal oscillates between minus one and one, and the state varables are not constant. Examples of this behaviour can be found in Fig. 3.29.

Since a Sigma-Delta modulator reaches no stable operating point, all Sigma-Delta modulators are **unstable**, according to the common definition of a stable system [1].

In this work, the following alternative definition of a stable Sigma-Delta modulator is used: a Sigma-Delta modulator is **stable** when for a bounded input signal, the state variables are bounded. It is shown in this section that the circuit of Fig. 4.1. is only conditionally stable. For some values of the input signals x and for some initial conditions of the state variables v_i, the state variables describe a nonlinear limit cycle with a large frequency outside the signal band. Examples will be given further on. For other combinations of input signals and initial conditions, the state variables grow without bound. This is comparable with the exponential growth of the state variables in an unstable linear system.

Because of the nonlinear quantiser characteristic, the stability of the circuit of Fig. 4.1. cannot be investigated with the well-known root locus technique or with the modified Schur-Cohn criterium [1], valid for linear systems only. The describing function technique [2] is not appropriate neither to describe a Sigma-Delta modulator. Hence, a rigourous mathematical stability analysis for this nonlinear circuit is not easy because no standard mathematical techniques can be used.

In this text, the stability conditions for the Sigma-Delta modulator loop are investigated in a pragmatic way. With the computer simulation routine described in chapter 3, section 3.7, the evolution in the time of the state variables is followed for a large number of different input signals, initial conditions and loop parameters. In this way, it can be verified under which conditions the modulator loop is stable. Although this approach yields no closed mathematical proof for the stability criteria formulated below, it allows to obtain enough insight in the modulator behaviour to design stable Sigma-Delta modulators.

In order to illustrate the stability problem in higher-order modulators, the transient behaviour of First- and a Second-order modulators is discussed first. Afterwards, the evolution of the limit cycles in higher-order modulators is compared with this behaviour.

4.2.a. The transient behaviour of a First-order modulator

Consider the First-order Sigma-Delta modulator of Fig. 4.2.a. and assume for a moment that the integrator gain equals one. This circuit can be described with a nonlinear difference equation:

$$v(k+1) = v(k) + [x(k) - y(k) \cdot V_{ref}] \qquad (4.1.a)$$

with
$$y(k) = \text{sign}(v(k)) \qquad (4.1.b)$$

In these expressions, x(k) is the input signal, y(k) is the modulator PDM output and v(k) is the integrator output voltage. With computer simulations, the signal v(k) can be analysed for different input values x and for different initial conditions of v(k).

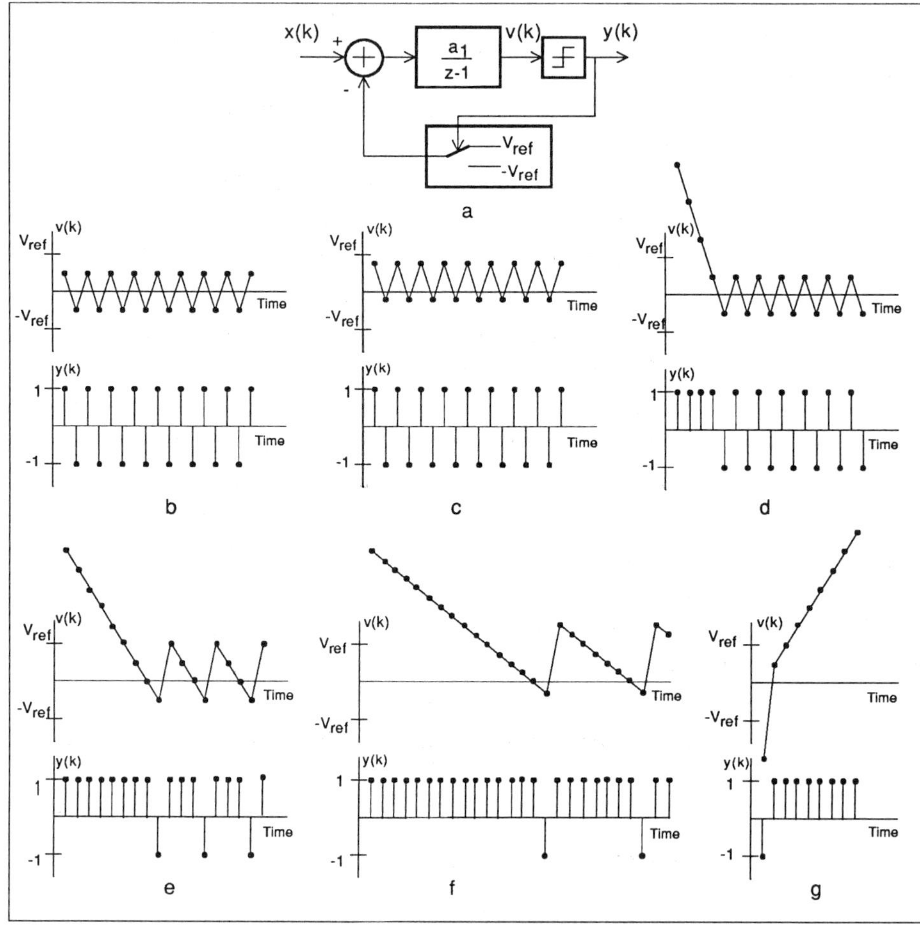

Fig. 4.2:
a) The basic schematic of a First-order modulator
b) The state variable versus time for x=0 and v(k) initially set to 0.5.V_{ref}
c) idem for v(k) initially set to 0.8.V_{ref}
d) idem for v(k) initially equal to 3.5.V_{ref}
e) idem for v(k) initially equal to 3.5.V_{ref} and x=0.5.V_{ref}
f) idem for v(k) initially set to 3.5.V_{ref} and x=0.75.V_{ref}
g) idem for v(k) initially set to -2.V_{ref} and x=1.5.V_{ref}

Examine first the situation where the input signal x is zero and where the initial value of v(k) equals $0.5.V_{ref}$. It can be easily verified that the comparator output y(k) and the integrator output voltage v(k) are as depicted in Fig. 4.2.b. v(k) oscillates between $-0.5.V_{ref}$ and $+0.5.V_{ref}$. The circuit describes a nonlinear limit cycle with a period of two clock cycles. Since the frequency of this oscillation is larger than the lowpass filter frequency $f_d/2$ (see Fig. 4.1.), there is no noise power inside the signal band. This nonlinear oscillation causes no degradation of the signal quality and can therefore be tolerated. Since the amplitude of the state variable is bounded, this is a stable oscillation.

Consider now the situation where the initial integrator output voltage equals $0.8.V_{ref}$. The corresponding signals v(k) and y(k) are depicted in Fig. 4.2.c. v(k) oscillates between $0.8.V_{ref}$ and $-0.2.V_{ref}$. Due to the different initial condition, the values of the state variable v(k) are different from those in Fig. 4.2.b. However, unlike in stable linear systems, this difference is not damped out when the time increases. Note that the output signal y(k) in Fig. 4.2.c. is identical to that in Fig. 4.2.b.

In Fig. 4.2.d, the situation is depicted for an initial condition of v(k) equal to $3.5.V_{ref}$. The input signal x still equals zero. After an initial transient of three clock cycles, the state variable is decreased below V_{ref}. The output starts to oscillate with a period of two clock cycles. In a similar way, when the initial condition is below -Vref, v(k) increases first up to a value higher than -Vref and then it oscillates with a period of two clock cycles.

From these three examples, it can be concluded that, whatever the initial condition of the state variable is, after a transient behaviour, the output starts to oscillate at half of the clock frequency and the state variable remains between -Vref and $+V_{ref}$. During this oscillation, v(k) is bounded between -Vref and $+V_{ref}$ and hence, the circuit is stable.

In order to further characterise the First-order modulator, consider the situation where a non-zero DC input signal x (between -Vref and $+V_{ref}$) is applied to the modulator input and where the initial value of v(k) is still equal to $3.5.V_{ref}$. Two examples are shown in Figs. 4.2.e. and f. After a transient, the output y(k) starts to oscillate. The oscillation frequency is a function of the input signal x. For a DC input of, for instance, $0.5.V_{ref}$, the oscillation period equals four clock cycles (see Fig. 4.2.e). For x equal to $0.75.V_{ref}$, the limit cycle takes 8 clock periods (see Fig. 4.2.f.). Especially for input voltages close to $+V_{ref}$ or -Vref, the oscillation frequency can be much smaller than the clock frequency f_s, resulting in a large amount of quantisation noise power at low frequencies in the signal band. This noise power will pass the lowpass filter in Fig. 4.1. and appear at the ADC output. This is the pattern noise, described in detail in chapter 3, section 3.4.a.

After the initial transient in Figs. 4.2.e and f, the amplitude of v(k) is bounded. This can be verified mathematically from expressions (4.1): when x is between -Vref and $+V_{ref}$, v(k+1) is given by

```
v(k+1)  =  v(k)+x-Vref  >  -Vref+x  >  -2.V_ref   when v(k)  > 0      (4.2.a)
v(k+1)  =  v(k)+x+V_ref  <   V_ref+x  <   2.V_ref   when v(k)  < 0:    (4.2.b)
```

The state variable cannot grow without bound. This oscillation is stable.

As a last example, consider the situation where the Sigma-Delta modulator is overloaded: an input signal x larger than $+V_{ref}$ (or similarly: smaller than -Vref) is applied to the modulator input. The corresponding signal v(k) at the integrator output is depicted in Fig. 4.2.g. The integrator output voltage increases without limit. This can be verified mathematically from expressions (4.1): when x is larger than V_{ref}, v(k+1) is larger than v(k). The modulator is unstable. However, when the overload input voltage is removed (i.e. the input is decreased to a value between $+V_{ref}$ and -Vref), v(t) will decrease back to a value between $+2.V_{ref}$ and $-2.V_{ref}$ and the modulator will repeat its normal limit cycle.

From these six examples, it can be concluded that the First-order modulator is stable, except when the modulator input is overloaded or in the initial condition, at power-up. The state variable v(k) is limited between $-2.V_{ref}$ and $+2.V_{ref}$. After the power-up or after an overload, the state variable will first evoluate to a value between $-2.V_{ref}$ and $+2.V_{ref}$ and then start a stable oscillation. This dynamic behaviour is well suited for a Sigma-Delta modulator.

4.2.b. The transient behaviour of a Second-order modulator

In Fig. 4.3.a, a Second-order Sigma-Delta modulator is depicted. This circuit contains two state variables, v(k) and w(k). The circuit operation is a function of three design variables, a_1, a_2 and b_2. In a practical design, these variables are determined by capacitor or resistor values.

Fig. 4.3.b. shows the same circuit, but now with the feedback factor b_2 and the gain a_2 of the second integrator equal to one. This circuit only differs from the previous one in the fact that the internal signals v(k) and w(k) are scaled.

The comparator in Fig. 4.3.b. is sensitive to the sign of its input signal, but not to the magnitude. Hence, the gain after the second integrator can be set equal to one without influencing the behaviour of the modulator loop.

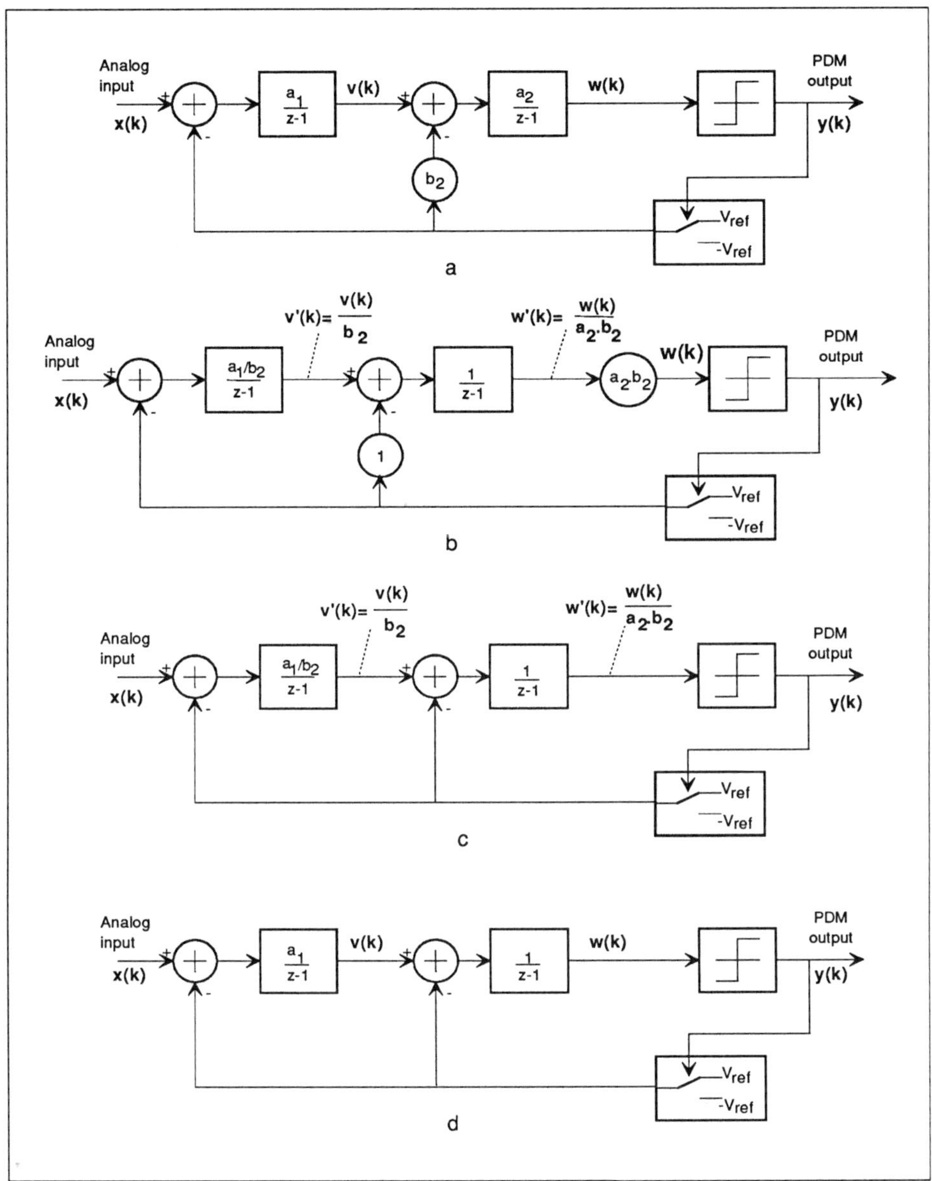

Fig. 4.3: Different implementations of the same Second-order Sigma-Delta modulator

The circuit of Fig. 4.3.c. is equivalent to the two previous ones. This implies that the parameters a_2 and b_2 can be set equal to one in Fig. 4.3.a. without loosing generality. There is only one relevant circuit parameter (a_1) left.

From now on, it will be assumed that a_2 and b_2 are equal to one in Fig. 4.3.a, which corresponds to the circuit of Fig. 4.3.d.

In the same way as for a First-order modulator, the evolution of the state variables in Fig. 4.3.d. can be investigated for different input signals, for different state variable initial conditions and for different values of the parameter a_1.

In a first attempt to characterise this Second-order modulator, the loop is simulated for different values of the parameter a_1, with the input signal x equal to zero, with v(k) initially set to V_{ref} and with w(k) initially equal to zero. It can be verified with the simulation routine described in section 3.7. that for a_1-values between zero and 0.8, the oscillation of the state variables is damped out, resulting in a stable limit cycle. A simulated example is illustrated in Fig. 4.4.b. for a_1 equal to 0.79.

For a_1 larger or equal than 0.8, the state variables grow without bound and the loop is unstable. This is illustrated in Fig. 4.4.a. for a_1 equal to 0.8.

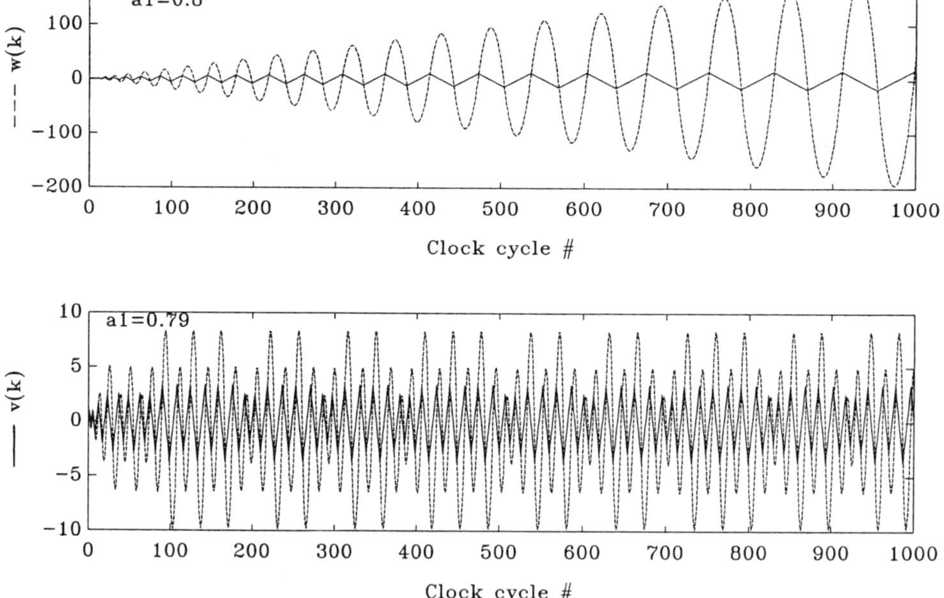

Fig. 4.4: The evolution of the state variables in a Second-Order modulator. x equals zero. v(k) and w(k) are initially set to V_{ref} and zero respectively.
a) a_1=0.8 and b) a_1=0.79

These simulations allow to conclude that Second-order Sigma-Delta modulators with a_1-values larger than 0.8 cannot be used in practice.

In the same way, the state variables can be simulated for non-zero input signals and for different initial conditions. We restrict ourselves to modulators with a_1 smaller than 0.8. For input signals below -Vref or higher than $+V_{ref}$, the loop is unstable. For input signals between $+V_{ref}$ and -Vref, the modulator loop reaches a stable limit cycle after an initial transient, regardless the value of a_1. Two examples are shown in Fig. 4.5.

The amplitude of the state variables are functions of the input signal value. Especially for input voltages close to $+V_{ref}$ or -Vref, the amplitude of w(k) is very large. This can be noticed in Fig. 4.5.b: for x equal to $0.9.V_{ref}$, w(k) oscillates at a low frequency with a maximum value of $50.V_{ref}$.

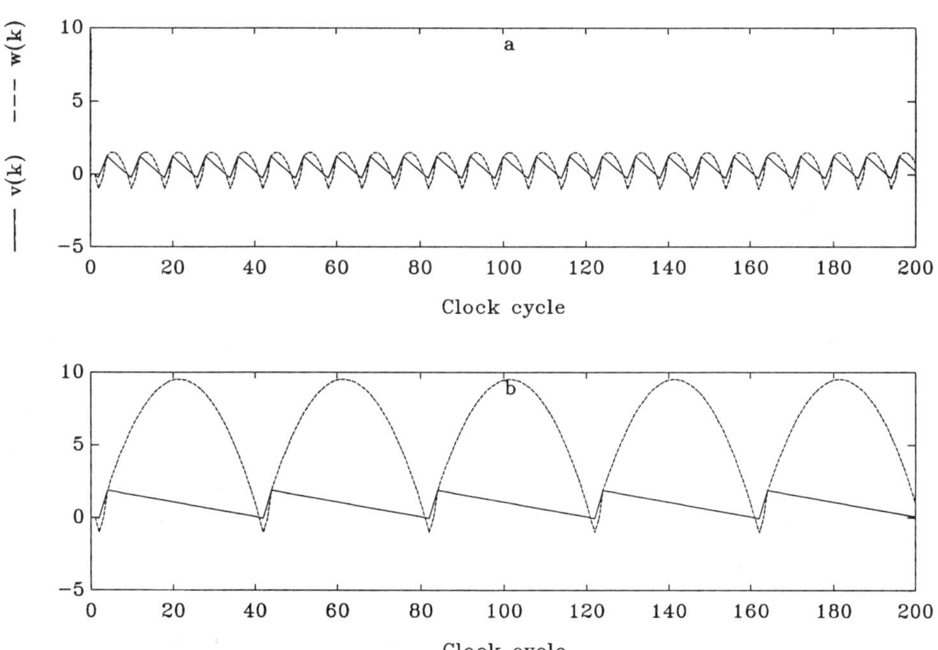

Fig. 4.5: The state variables of a Second-order modulator
for v(k) and w(k) initially set to zero, for a_1 equal to 0.5
and for a) x equal to $0.5.V_{ref}$ and b) x equal to $0.9.V_{ref}$

In Fig. 3.27.c, the state variable amplitudes are plotted versus the input signal x. As can be seen from this figure, the amplitude of w(k) grows assymptotically when x approaches $+V_{ref}$ or -Vref. In a practical design, the voltage w(k) will clip to the supply voltages for large input signals and the modulator will not operate properly. As a result, the in-band quantisation noise increases abruptly for input signals close to $+V_{ref}$ or -Vref. This behaviour can be noticed in Fig. 3.31. Where for a First-order modulator the useful input range extends from -Vref to $+V_{ref}$, the useful input range for a Second-order modulator is approximately between $-0.9.V_{ref}$ and $+0.9.V_{ref}$, depending on the practical realisation. For input signals outside this range, the modulator will overload and one or more of the state variables will clip to the supply voltages.

From these computer simulations, it can be concluded that a Second-order Sigma-Delta modulator is stable provided that the gain a_1 in Fig. 4.3.d. is between zero and 0.8 and that the input signal is approximately between $-0.9.V_{ref}$ and $+0.9.V_{ref}$. The first condition is equivalent to a ratio b_2/a_1 in Fig. 4.3.c, larger than 1.25. This conclusion is in agreement with [3].

When the modulator is overloaded (i.e. an input signal larger than $+V_{ref}$ or smaller than -Vref is applied), the state variables grow without bound. When afterwards, the input voltage is changed to a value between $+V_{ref}$ and -Vref, the state variables converge back to their limit cycle and the modulator will repeat its steady-state regime.

This dynamic behaviour is also well suited for a Sigma-Delta ADC.

4.2.c. The stability of a Fourth-order modulator

Compared with a Third-order Sigma-Delta modulator, only one extra integrator is required to construct a Fourth-order modulator. But for the latter, the speed-accuracy trade off is more advantageous. The stability problems of a Third-order Sigma-Delta modulator are similar to those of a Fourth-order modulator. When these stability problems can be solved, there is no reason to prefer a Third-order modulator over a Fourth-order one. Therefore, we skip the Third-order modulator and derive stability conditions for a Fourth-order Sigma-Delta modulator.

In Fig. 4.6, the schematic of a Fourth-order modulator is depicted. Again, the feedback factors and the last integrator gain can be set equal to one without loosing generality. This circuit contains four state variables, t(k), u(k), v(k) and w(k). The circuit operation is a function of three design parameters, a_1, a_2 and a_3.

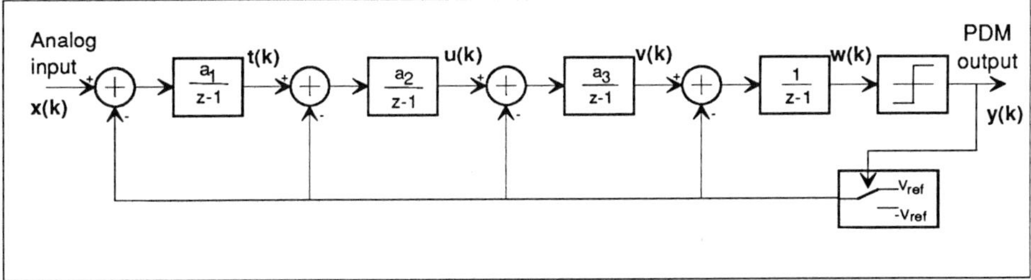

*Fig. 4.6: The principle schematic of a Fourth-order
modulator*

In a first attempt to obtain insight in the dynamic behaviour of this nonlinear system, the evolution in time of the four state variables is simulated with the routine of section 3.7. for one particular combination of the design parameters (a_1=0.15, a_2=0.15, a_3=0.3), for different input values x, and **with the state variables initially set to zero.** For small x-values, between -0.58.V_{ref} and +0.58.V_{ref}, the simulations show that the state variables are bounded. For x-values outside this range, the state variables diverge. This is illustrated in Fig. 4.7: for x equal to 0.57.V_{ref}, the the state variables are bounded after an initial transient. For x equal to 0.6.V_{ref}, the state variables grow without limit. Clearly, this circuit cannot be used for input signals larger than +0.58.V_{ref} or smaller than -0.58.V_{ref}.

These simulations can be repeated for other integrator gain values. It turns out that for each set of integrator gains a_1, a_2 and a_3, there is a **stable input range R**. For x-values between -R and R, the state variables are bounded while for input signals outside this range, one or more of the state variables grow without limit.

In Fig. 4.8, this stable input range R is plotted as a function of the integrator gains. E.g. for a_1, a_2 and a_3 equal to 0.15, 0.15 and 0.3 respectively, R is equal to 0.58.V_{ref}. This implies that the circuit is stable for input signals x between -0.58.V_{ref} and +0.58.V_{ref}. For a_1, a_2 and a_3 equal to 0.3, 0.4 and 0.5 respectively, the stability range R is zero. For this combination of integrator gains, the oscillation is unstable for all input signal values.

With a well-considered choice of the integrator gains, a large stable input range R can be obtained. For instance for a_1 and a_2 equal to 0.05 and for a_3 equal to 0.3, the modulator is stable for input signals between -0.83.V_{ref} and +0.83.V_{ref}. However, capacitor or resistor ratios as large as 20 to 1 are required to realise an integrator gain of 0.05. A design with a_1, a_2 and a_3 equal to 0.1, 0.1 and 0.4 respectively is a better option from this point of view. A stable input range of 0.75.V_{ref} is obtained with component ratios of only 10 to 1.

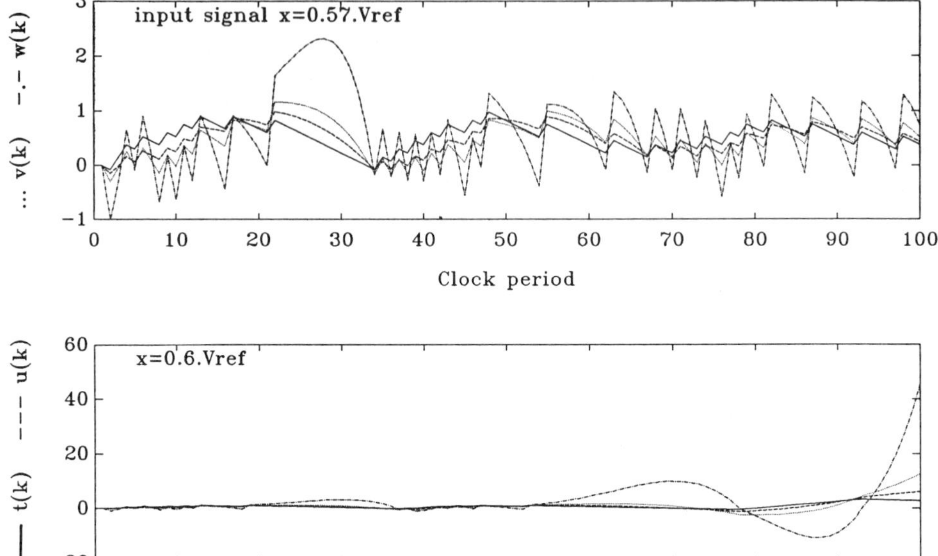

*Fig. 4.7: The state variables versus time for $a_1 = a_2 = 0.15$,
$a_3 = 0.3$, for initial conditions equal to zero and with two
different input signals.*

In Fig. 4.9, the amplitudes of the state variables are plotted versus the input signal value x. For x between $-0.75.V_{ref}$ and $+0.75.V_{ref}$, the amplitudes of t(k), u(k), v(k) and w(k) are below $0.9.V_{ref}$, $0.9.V_{ref}$, $1.25.V_{ref}$ and $2.5.V_{ref}$ respectively. With a proper scaling of the signals at the integrator outputs, it can be assured that the state variables are always well in between the two supply voltages. For |x| larger than $0.75.V_{ref}$, the amplitudes of the state variables abruptly increase without bound.

At this point, it can be concluded that the Fourth-order modulator with integrator gains a_1, a_2 and a_3 equal to 0.1, 0.1 and 0.4 respectively is stable, provided that the input signal remains between $-0.75.V_{ref}$ and $+0.75.V_{ref}$ and that the state variables are initially set equal to zero. Note again that for all the simulations described up till now in section 4.2.c, the initial conditions of the state variables were equal to zero.

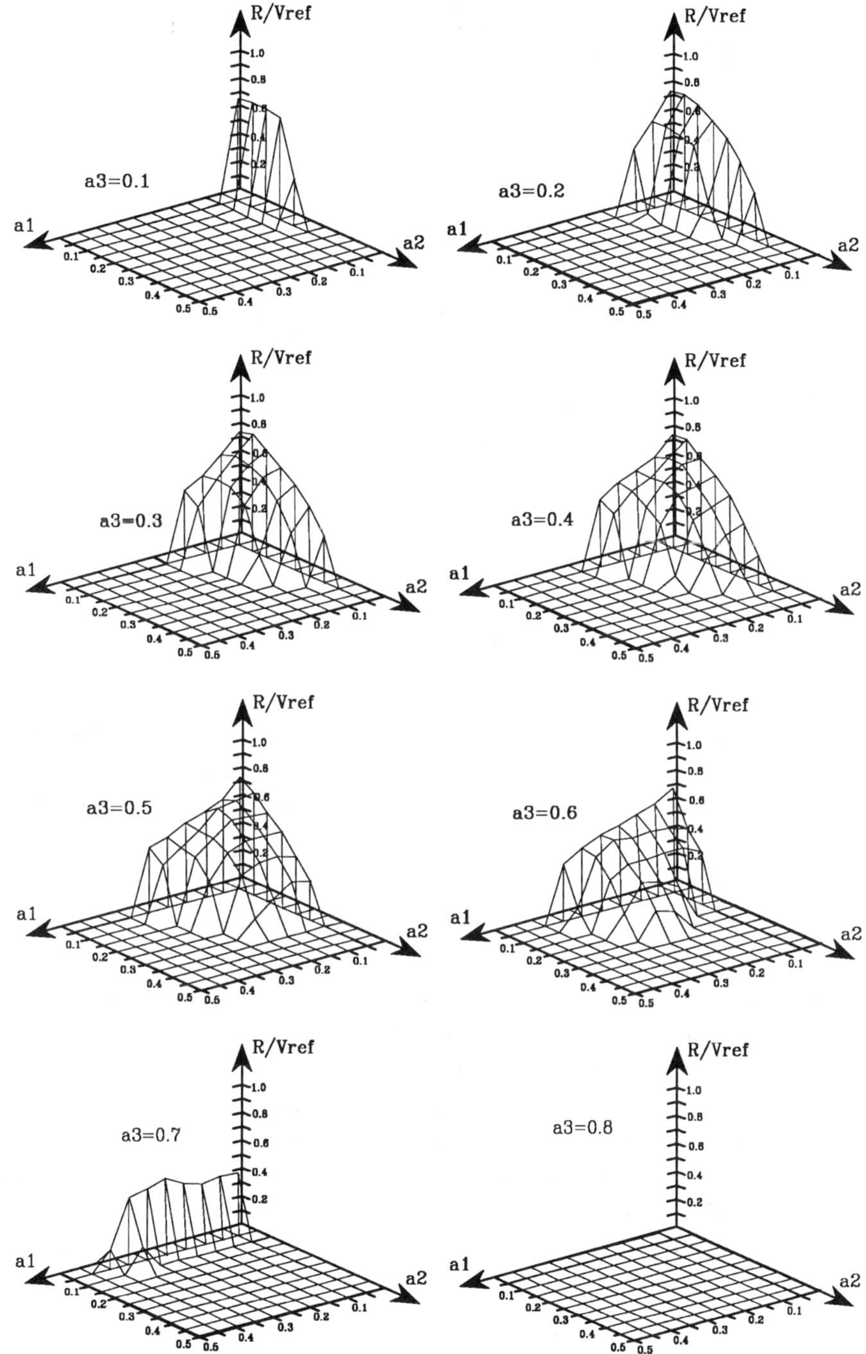

Fig. 4.8: The stable input range R as a function of the design variables.

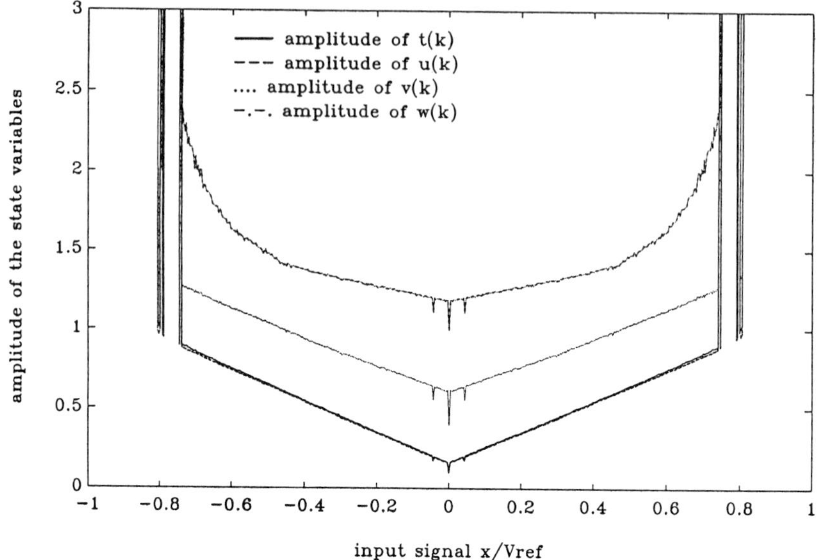

*Fig. 4.9: The amplitudes of the state variables versus the
input signal for a_1=0.1, a_2=0.1, a_3=0.4 and for initial
conditions equal to zero.*

In order to investigate the modulator behaviour for non-zero initial conditions, a simulation is performed with x equal to zero and with all the state variables initially set to $3.V_{ref}$. The simulation results are shown in Fig. 4.10. As can be seen, v(k) and w(k) grow without bound and hence, the loop is unstable.

This is a troublesome complication: even when the integrator gains are optimised towards a large stable input range R and even when the input signal x is inside this range, the loop is unstable for some non-zero initial conditions.

Another problem is the loop behaviour after an overload. When an input signal outside the stable input range R is applied to the modulator input, the state variables grow without bound as explained before (see Fig. 4.7.b.) When afterwards, the input signal is reduced to a value inside the stable range, the state variables have already grown to unpredictable values. It is not guaranteed that for these non-zero initial conditions, the state variables will converge back to their normal values. Hence, it is not sure that after an overload, the modulator will repeat its limit cycle.

These stability problems have delayed the developement of higher-order modulators for several years.

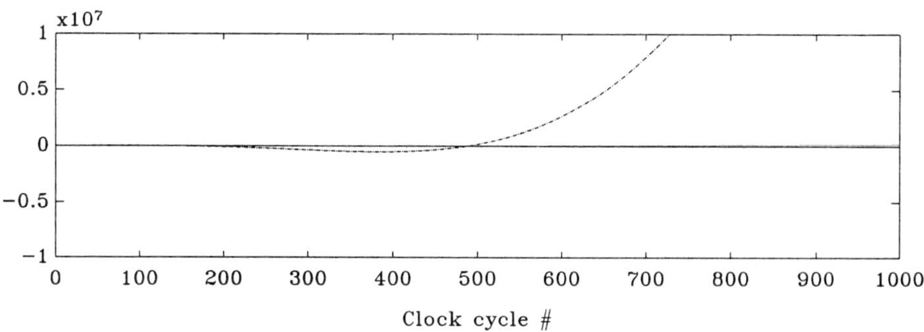

Fig. 4.10: The state variables versus time for a_1=0.1, a_2=0.1, a_3=0.4, for x=0 and for all the state variables initially set to $3.V_{ref}$

As a first solution to overcome the loop instability [4], the state variables can be initialised equal to zero at the power-up. In this way, non-zero initial conditions are prevented. When the input signal is then maintained inside the stable input range R, the state variables are bounded. According to our opinion, this is not a reliable solution: a spike on a power-supply voltage or a short power distribution failure can modify the state variables abruptly. It is not guaranteed that after such an event, the state variables will converge back to their limit cycle.

As a further improvement, the state variables can be continuously monitored. This is shown in Fig. 4.11. Under the normal stable operation of the loop, the state variables t(k) and u(k) are always between -0.9.V_{ref} and +0.9.V_{ref}. Similarly, the amplitudes of v(k) and w(k) are below 1.25.V_{ref} resp. 2.5.V_{ref}. When the loop is unstable, the state variables grow without bound. With additional detector circuits, it can be dectected when the state variables grow beyond their limits. When such an unstabe behaviour is detected, the loop is forced into a stable state by resetting the state variables to zero [4].

Note again that such an unstability can only occur at the initial power-up, at an input overload (i.e. when an input signal x with magnitude larger than 0.75.V_{ref} is applied) or at a short power distribution failure, but not during the normal operation of the ADC. The detectors of Fig. 4.11. do not operate under normal operation and therefore they cause no degradion of the ADC output signal quality.

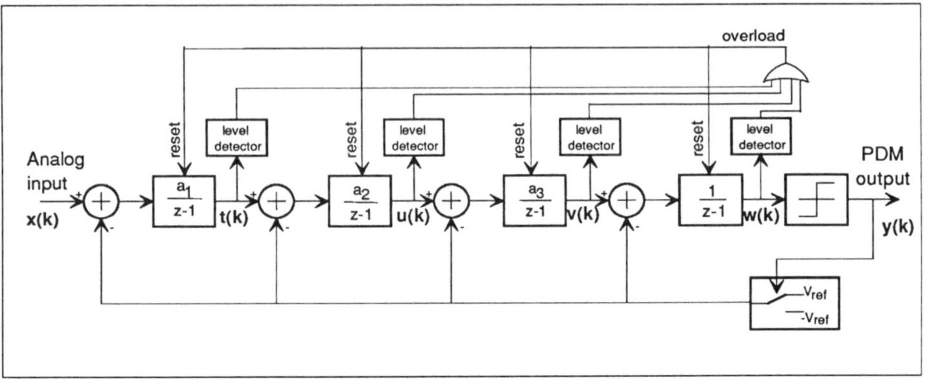

*Fig. 4.11: a stable Fourth-order Sigma-Delta modulator
with level detectors*

As a last solution to overcome the stability problems mentioned above, it is shown here that an unstable behaviour of the circuit of Fig. 4.6. can be prevented with a well-considered scaling of the internal signals at the integrator outputs.

An unlimited growth of the state variables as shown in Fig. 4.10. is not realistic: in a practical design, the state variables cannot grow without bound since they are limited by the supply voltages. This effect can be taken into account in the computer simulations with a more elaborate macromodel, with a clipper after each integrator. This macromodel is illustrated in Fig. 4.12.a.

As stated before, a Fourth-order Sigma-Delta modulator is only conditionally stable. Even for small input signals, the loop is unstable for some non-zero state variable initial conditions. When the clipper levels in Fig. 4.12.a. are determined in such a way that the loop is stable for all possible initial conditions that are situated between the clipper levels, initial conditions that cause instability become impossible and the stability of the loop is guaranteed.

Compared with the Fourt-order modulator of Fig. 4.6, the behaviour of the circuit of Fig. 4.12.a. depends on four additional design variables, namely the clipping levels of the four integrators. No mathematical technique is available to analyse this circuit in ana analytical way. Therefore, in this work, the stability of the Fourth-order modulator of Fig. 4.12.a. is analysed with behavioural simulations, with the following routine:

```
set a₁ to 0.1, a₂ to 0.1, a₃ to 0.4, a₄ to 1,
    b₁ to 1, b₂ to 1, b₃ to 1, b₄ to 1
for t_clip = 0.9 to 10  step 0.05
  for u_clip = 0.9 to 10  step 0.05
    for v_clip = 1.25 to 10  step 0.05
      for w_clip = 2.5 to 10  step 0.05
          for t_init = -tclip to tclip step 0.05
            for u_init = -uclip to uclip step 0.05
              for v_init = -vclip to vclip step 0.05
                for w_init = -wclip to wclip step 0.05
                  for x = -0.75 to 0.75 step 0.05


                  " Simulate the modulator of Fig. 4.12.b. with
                    the routine described in section 3.7."

                  " When some of the state veriables remain
                    clipping after 10000 clock periods:
                            the loop is unstable
                    else   the loop is stable "


                        if unstable goto ───────────────┐
                      end                                │
                    end                                  │
                  end                                    │
                end                                      │
              end                                        │
                                                         │
        " a set of limiter levels is found               │
          which guarantees absolute stability "          │
                                                         │
              <───────────────────────────────────────────┘
      end
    end
  end
end
```

t_{clip}, u_{clip}, v_{clip} and w_{clip} are the clipper levels for the signals t, u, v, and w respectively. t_{init}, u_{init}, v_{init} and w_{init} are the state variable initial conditions.

These computer simulations revealed that the circuit of Fig. 4.12.a. is stable provided that four conditions are met:

1)   The input signal x has to be inside the stable input range, between $-0.75.V_{ref}$ and $+0.75.V_{ref}$.

2)   With a clipper after the third integrator, v(k) is clipped between v_{clip} and $-v_{clip}$. The clipper level v_{clip} has to be smaller than $2.5.V_{ref}$.

3)   In a similar way, w(k) is clipped between w_{clip} and $-w_{clip}$. This clipper voltage w_{clip} has to be smaller than $5.V_{ref}$.

4)   No clippers are required for the signals t(k) and u(k).

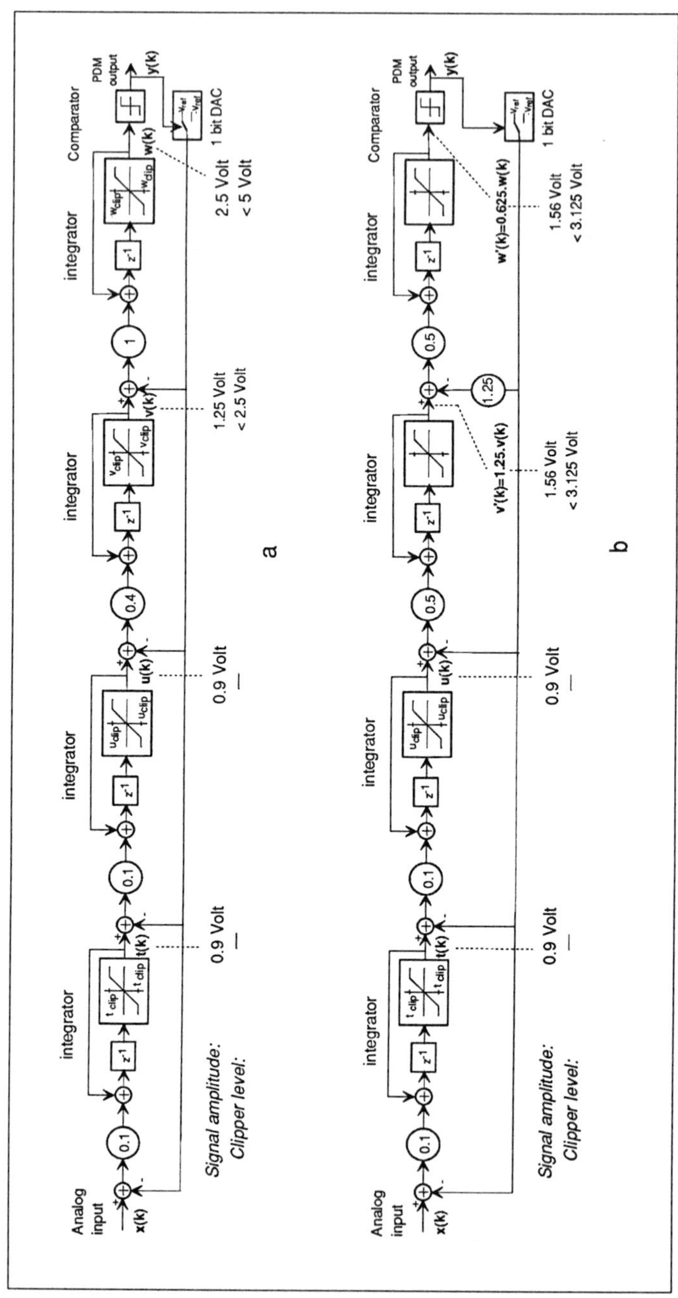

Fig. 4.12: macromodels of the Fourth-order modulator,
including the clipping effects of the supply voltages
a) with the feedback factors and the
last integrator gain equal to one
b) with scaled internal signals

In order to illustrate these conditions, some simulated examples are shown in Fig. 4.13. For the curves of Fig. 4.13.a to c, the four conditions are met and the limit cycle converges after about 800 clock periods. In Fig. 4.13.d, w(k) is clipped to $6.V_{ref}$ and condition 3) is not met. In Fig. 4.13.e, condition 2) is violated. For these two last examples, the state variables clip to the supply voltages, even in the steady-state regime after the initial transient. Figs 4.13.d. and e. clearly illustrate that the clippers are essential for the stability of the loop.

It is quite simple to implement the clippers of Fig. 4.12.a. Consider for instance the situation where the reference voltage is equal to 1 Volt and where the supply voltages are +2.5 Volt and -2.5 Volt. The maximum signal amplitudes on the different nodes, obtained from Fig. 4.10, are indicated in Fig. 4.12.a. The maximum values for the clipper levels v_{clip} and w_{clip}, according to conditions 2) and 3), are also shown.

In Fig. 4.12.b, the same circuit is depicted with scaled internal signals. Since the gain of the third integrator is a factor 1.25 larger than in Fig. 4.12.a, its output is equal to 1.25.v(k). The last feedback factor is scaled accordingly. The last integrator gain is reduced to 0.5 and its output voltage therefore equals 0.625.w(k). As a result, the maximum amplitude of these two signals equals 1.56 Volt. This is well in between the two supply voltages. Conditions 2) and 3) are met when the two clipper levels are smaller than 3.125 Volt. Since in practice, the integrator output voltages are clipped by the supply voltages (+2.5 Volt and -2.5 Volt), this condition is met and the loop is stable.

For other values of the supply voltages or the reference voltage, other scaling factors have to be used. Note that the signals v'(k) and w'(k) should only clip to the supply voltages after the initial power-up or after an input overload, but not during the steady-state regime.

4.2.d. Conclusions

From the large amount of computer simulations described above, the following conclusions are retrieved:

-   A First-order Sigma-Delta modulator describes a stable limit cycle, regardless the initial condition of the state variable, provided that the input voltage is between -Vref and $+V_{ref}$. However, for input voltages close to V_{ref} or -Vref or close to zero, the oscillation frequency is small. This results in a large amount of in-band pattern noise power.
    After an overload, the state variable converges back to its limit cycle.
-   For a second-order modulator with b_2/a_1 larger than 1.25 (see Fig. 4.3.a), the state variables describe a stable limit cycle, regardless their initial conditions, provided

that the input signal is between $-0.9.V_{ref}$ and $+0.9.V_{ref}$. For larger input signals, the output of the last integrator will clip to the supply voltages.

After an overload, the state variables will return to their limit cycle.

- A Fourth-order modulator is stable for some initial conditions of the state variables and unstable for others. After an overload, the state variables will not necessary return to their stable limit cycle.

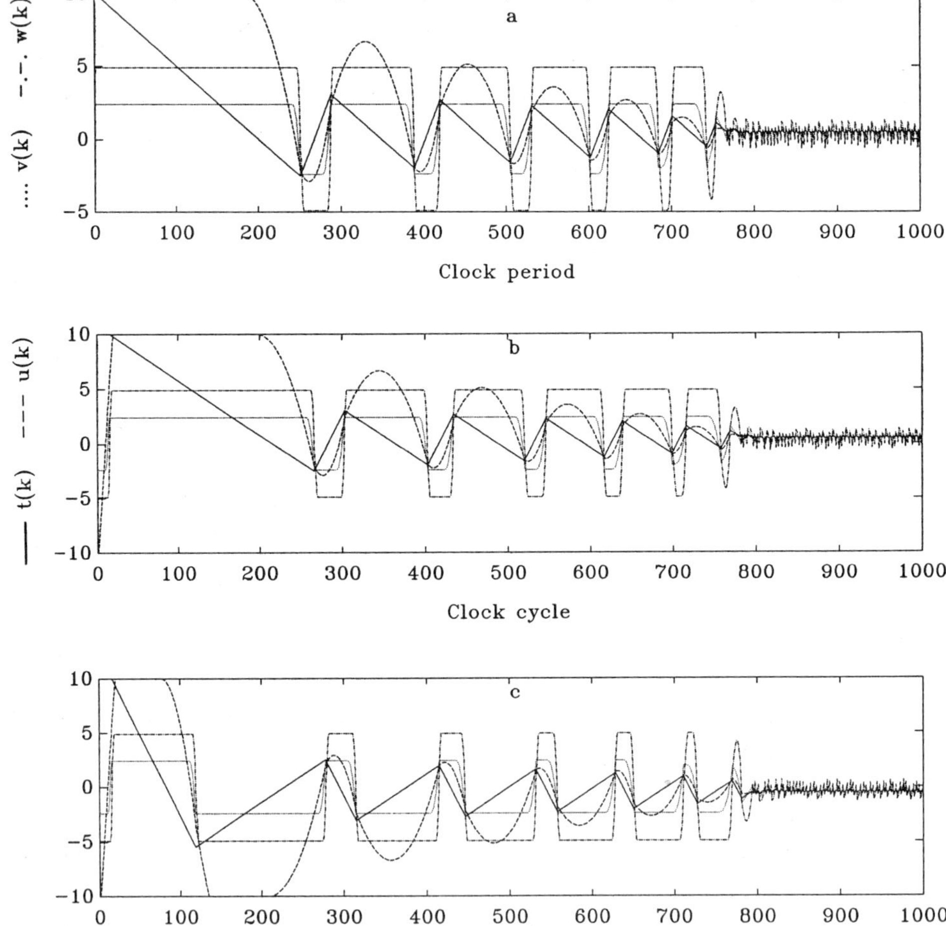

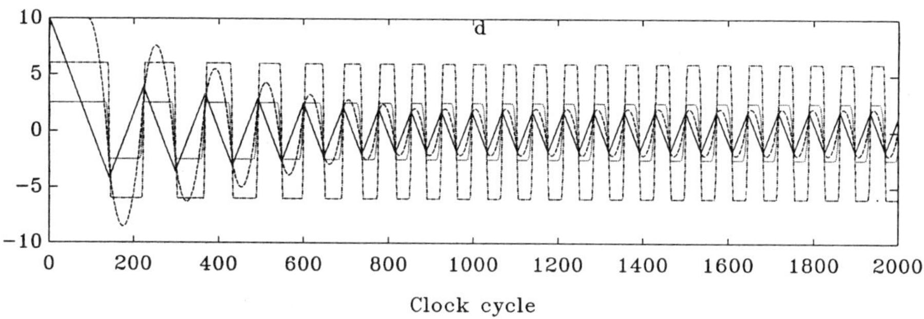

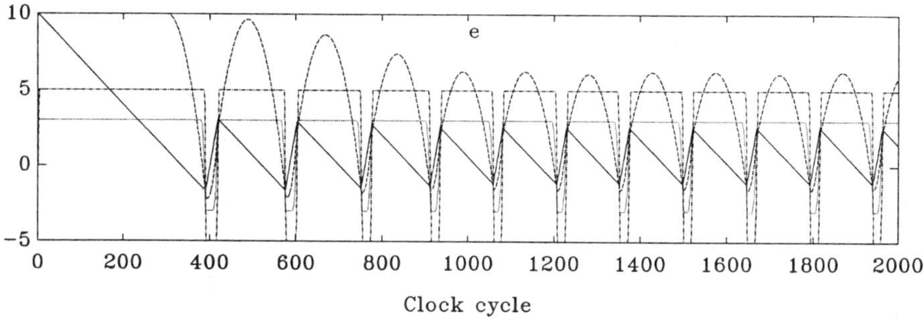

Fig. 4.13: The evolution of the state variables in time for $a_1=a_2=0.1$, $a_3=0.4$ and for

> *a) t(k) and u(k) initially set to $10.V_{ref}$, v(k) clipped at $2.4.V_{ref}$ and w(k) clipped at $4.9.V_{ref}$. x is $0.5.V_{ref}$*
> *b) idem but w(k) initially set to $-2.4.V_{ref}$*
> *c) idem but $x = -0.5.V_{ref}$*
> *d) idem but x=0 and w(k) clipped at $6.V_{ref}$*
> *e) idem as a) but x=0 and v(k) clipped at $3.V_{ref}$*

- With a well-considered choice of the loop parameters and with clippers on the two last integrator outputs, it can be assured that the state variables describe a stable limit cycle for input signals between $-0.75.V_{ref}$ and $+0.75.V_{ref}$. After an overload, the state variables return to this limit cycle. With a proper scaling of the integrator output signals, no additional citcuitry is required for the clippers: the integrator outputs are clipped by the supply voltages.

- In the same way, it can be shown that the stability of a third-order modulator can be assured by proper scaling of the last integrator output voltage. In general, an n-th order modulator can be stabilised with clippers on the n−2 last integrator outputs.

These conclusions are derived from computer simulations. Although the simulations are extensively described, so that the reader can verify the results, it is impossible to check all the possible combinations of input signals and initial conditions exhaustively. These simulations therefore provide no formal proof for the conclusions presented above. Therefore, as a final proof, a practical realisation of the Fourth-order modulator is presented below, in section 4.4. Some practical design considerations are given first in section 4.3.

4.3. PRACTICAL DESIGN CONSIDERATIONS FOR SIGMA-DELTA ADCS

In Fig. 4.1, the principle schematic of a Sigma-Delta ADC is shown, as it is discussed in chapter 3. When taking only the quantisation noise into account, the SNR is expressed by formula (3.27). In a practical realisation however, the signal is further degraded by component nonidealities such as noise, device nonlinearities, integrator leakage, clock feedthrough or parasitic coupling between the digital and the analog sections of a chip.

The higher the dynamic range of an ADC, the more important these second-order effects are and the more difficult it is to translate the theoretical principle into silicon. In this section, the major second-order signal-degradation mechanisms are discussed. Wherever possible, design techniques are presented to reduce their influence.

4.3.a. The signal degradation due to integrator gain deviations or due to the sampling nonlinearity

Each of the integrators of Fig. 4.1. can be described with a difference equation:

$$v_i(k+1) = v_i(k) + a_i \cdot z_i(k) \tag{4.3}$$

where $v_i(k)$ and $v_i(k+1)$ are the i-th integrator output voltages on time k and k+1 respectively. $z_i(k)$ is the i-th integrator input voltage. The integrator gains a_i and the feedback factors b_i in Fig. 4.1. are determined by capacitor or resistor values. Due to component tolerances, these parameters can differ from their nominal values. The sensivity of the Sigma-Delta modulator for these tolerances is investigated in this section.

A First-order Sigma-Delta modulator (see Fig. 4.2.a.), contains only one integrator. Since the comparator is sensitive only to the sign and not to the magnitude of its input, the value of the integrator gain is arbitrary as long as it is positive.

For a second-order modulator with two integrators (see Fig. 4.3.a.), the gain of the last integrator is arbitrary. As shown in section 4.1, the stability condition requires that the ratio b_2/a_1 is larger than 1.25. Hence, when for instance this ratio is designed equal to two, a deviation of 38% on the parameters a_1 or b_2 can be tolerated.

For the Fourth-order modulator of Fig. 4.7.b, the gains a_1, a_2 and a_3 determine the stable input range R. Assumed that these gains have nominal values equal to 0.1, 0.1 and 0.4 respectively, tolerances of 25% cause a decrease of R with only 2% (see Fig. 4.8).

In general, it can be concluded that much larger integrator gain deviations can be tolerated in a Sigma-Delta modulator than for instance in a Switched-Capacitor filter. When the integrators are implemented with Switched-Capacitor techniques, large capacitor mismatches or incomplete charge transfers can be accepted.

The situation is worse when the first integrator gain is signal dependent. Assume for instance that the first integrator of Fig. 4.1. can be described with the following nonlinear difference equation:

$$v_1(k+1) = v_1(k) + a_{11}.z_1(k) + a_{12}.z_1(k)^2 + \ldots \qquad (4.4)$$

a_{11} represents the linear gain and a_{12} describes the second order integrator nonlinearity. This nonlinearity is due to the amplifier slew-rate [5], due to nonlinear switch resistances or due to capacitor nonlinearities. In appendix 4.A, it is pointed out that this nonlinearity causes a second harmonic distortion at the ADC output, equal to:

$$HD_2 = \frac{1}{2}.\frac{a_{12}}{a_{11}}.A_{in} \qquad (4.5)$$

where A_{in} is the ADC input signal amplitude. For instance for a 14 bit ADC with an input range of 1 Volt, the integrator nonlinearity a_{21}/a_{11} has to be below 0.01%/Volt. This illustrates that the linearity of the first integrator is extremely important for a high dynamic range ADC.

The nonlinearity of the second integrator of Fig. 4.1. can be described with an error voltage at its input. To calculate the equivalent error voltage at the ADC input, this error voltage has to be divided by the gain of the first integrator. Since this gain is large for in-band signals, the resulting input error voltage is small. Hence, the nonlinearity of the second, and also of the following integrators is less important.

4.3.b. Switched-Capacitor versus Continuous-Time integrators

The integrators and summators of Fig. 4.1. can be realised either with Switched-Capacitor techniques as shown in Fig. 4.14.a. or either with Continuous-Time circuits as depicted in Fig. 4.14.b. Since a large tolerance on the integrator gain can be accepted, process variations on the RC-products in Fig. 4.14.b. are not important. It is shown here that the Continuous-Time version is less suited for an ADC with a high sampling rate or with a high dynamic range.

When a single-pole model is assumed for the opamp in Fig. 4.14.a. and when its DC gain is assumed large, the integrator gain of the Switched-Capacitor realisation can in first-order by approximated by [7]:

$$a_i = \frac{C_1}{C_2} . [1 - \exp(-\frac{C_2}{C_1+C_2} . \frac{T_{\phi 2}}{\tau})]\qquad(4.6.a)$$

where $\tau = 1/2\pi GBW$ $\qquad\qquad(4.6.b)$

In these expressions, GBW is the opamp gainbandwidth, $T_{\phi 2}$ is the clock on-time and τ is the amplifier time constant. The non-zero switch on-resistances and the opamp slew rate are not considered in this approximation.

The Continuous-Time integrator gain can be calculated as

$$a_i = \frac{T_{\phi 2}}{R_1 C} . [1 - \frac{\tau}{R_1 C}]\qquad(4.7)$$

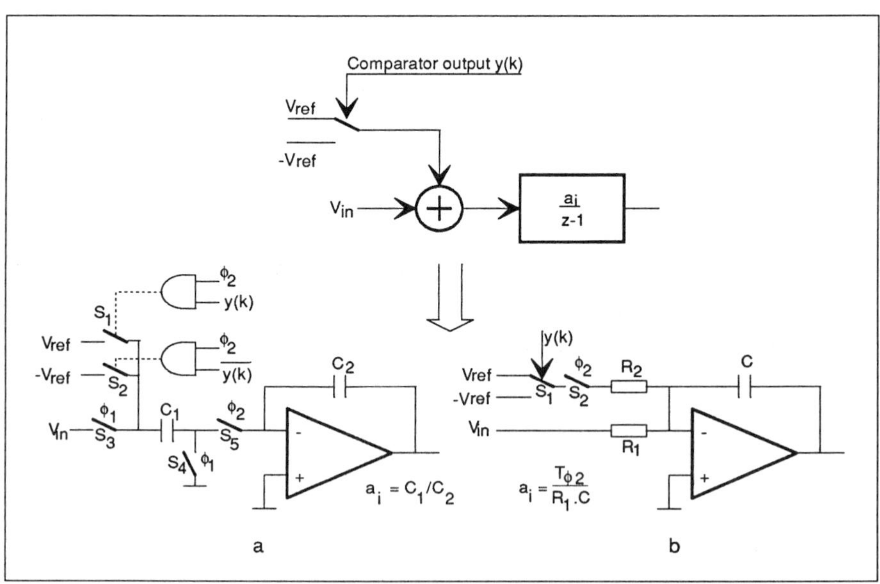

Fig. 4.14: a) Switched-Capacitor and
b) Continuous-Time integrators

In both expressions (4.6.a) and (4.7), the second term between the brackets represents a gain error due to the finite amplifier GBW. A nonlinearity in this term causes a nonlinear integrator gain, resulting in harmonic distortion. Hence, this term has to be small.

The Continuous-Time circuit of Fig. 4.14.b. is more sensitive for gainbandwidth limitations: for instance for a_i equal to one, a clock on-time $T_{\phi 2}$ of $1000.\tau$ is required to obtain an error of 0.1%. A Switched-Capacitor realialisation according to Fig. 4.14.a. requires a clock on-time of only $7.\tau$ for the same accuracy.

In most Sigma-Delta modulators, the finite opamp GBW is the main speed limiting factor. A modulator with Switched-Capacitor integrators can achieve higher sampling rates compared with a Continuous-Time realisation using the same amplifiers.

As a second drawback, Continuous-Time implementations are more sensitive to clock jitter than Switched-Capacitor realisations. For the circuit of Fig. 4.14.b, the amount of charge stored on the integrating capacitance during one clock cycle equals

$$Q(k) \; = \; (T_{\phi 2}{+}\Delta T) \,.\, (\frac{V_{in}}{R_1} \; - \; y(k)\,.\,\frac{V_{ref}}{R_2}) \qquad\qquad (4.8)$$

where ΔT represents the clock on-time error due to clock jitter. As can be seen, the clock jitter causes an error charge on the integrator capacitance. When ΔT is assumed random, with a white spectrum, the equivalent in-band input noise power can be calculated:

$$N \; = \; [\, \frac{\sigma(\Delta T)}{T_{\phi 2}} \,]^2 .\, \frac{V_{ref}^2}{2.OR} \qquad\qquad (4.9)$$

When $T_{\phi 2}$ is for instance equal to half of the clock period, the SNR equals:

$$SNR \; = \; -20.\log(\frac{\sqrt{OR}}{2f_s.\sigma(\Delta T)}) \;\; dB \qquad\qquad (4.10)$$

For the Switched-Capacitor realisation of Fig. 4.14.a, the charge transferred during one clock period is independent of the clock on-time. The aperture noise of this circuit is given by expression (3.51).

A comparison of expressions (4.10) and (3.51) reveals that the noise due to clock jitter is $20.\log(2.OR/\pi)$ worse for a Continuous-time realisation than for a Switched-Capacitor implementation. For instance for a clock jitter of 100 psec., for an OR of 100 and for a sampling frequency of 10 MHz, expression (4.10) yields a resolution of less than 12 bit. This explains why the Continuous-Time ADC realisations presented in the literature [8] [10] [11] are either slow or either achieve only 12 bit or less.

From now on, only Switched-Capacitor ADC realisations will be considered in this text.

4.3.c. The signal degradation due to integrator offset voltages

An integrator without external feedback is DC unstable because its offset voltage is integrated. The external feedback loop via the quantiser in Fig. 4.1. is necessary to stabilise the integrator operating points. When for instance the offset voltage of the second integrator is non-zero, the loop will build up a DC voltage at the first integrator output to compensate for this offset voltage. Hence, the offset voltage of the second and following integrators is not of major importance. It only affects the dynamic output range of the integrators. In the same way, the comparator offset voltage causes no signal degradation since it is compensated by a DC voltage, built up at the last integrator output.

The offset voltage of the first integrator is in series with the input signal. It causes an offset for the ADC.

4.3.d. The signal degradation due to integrator leakage

When calculating the quantisation noise and the SNR of the Sigma-Delta ADC depicted in Fig. 4.1, (see expressions (3.25) and (3.27)), it was assumed that the transfer function of each integrator has a pole at DC and a large gain for frequencies in the signal band. The in-band quantisation noise spectrum is suppressed by this large gain. This is illustrated in Fig. 4.15.a. and b. The ideal quantisation noise spectrum has an n-fold zero at DC.

In practice however, due to finite amplifier DC gains [7], the transfer function of each integrator is of the form:

$$H_i(z) = \frac{a_i}{z - (1-L_i)} \quad \text{with} \quad L_i = \frac{a_i}{A_{DC,i}} \tag{4.11}$$

L_i is the leakage of the i-th integrator, and $A_{DC,i}$ represents the i-th amplifier DC gain. Due to this leakage, the integrator poles are shifted from DC to frequencies given by:

$$f_{pi} \approx \frac{L_i}{2\pi}.f_s \quad i = 1..n \tag{4.12}$$

The integrator gain is now smaller at low frequencies, as shown in Fig. 4.15.c. The resulting quantisation noise power spectrum is shown in Fig. 4.15.d. The zeros are shifted from DC to the frequencies f_{pi} and the low-frequency quantisation noise is larger than for the ideal case.

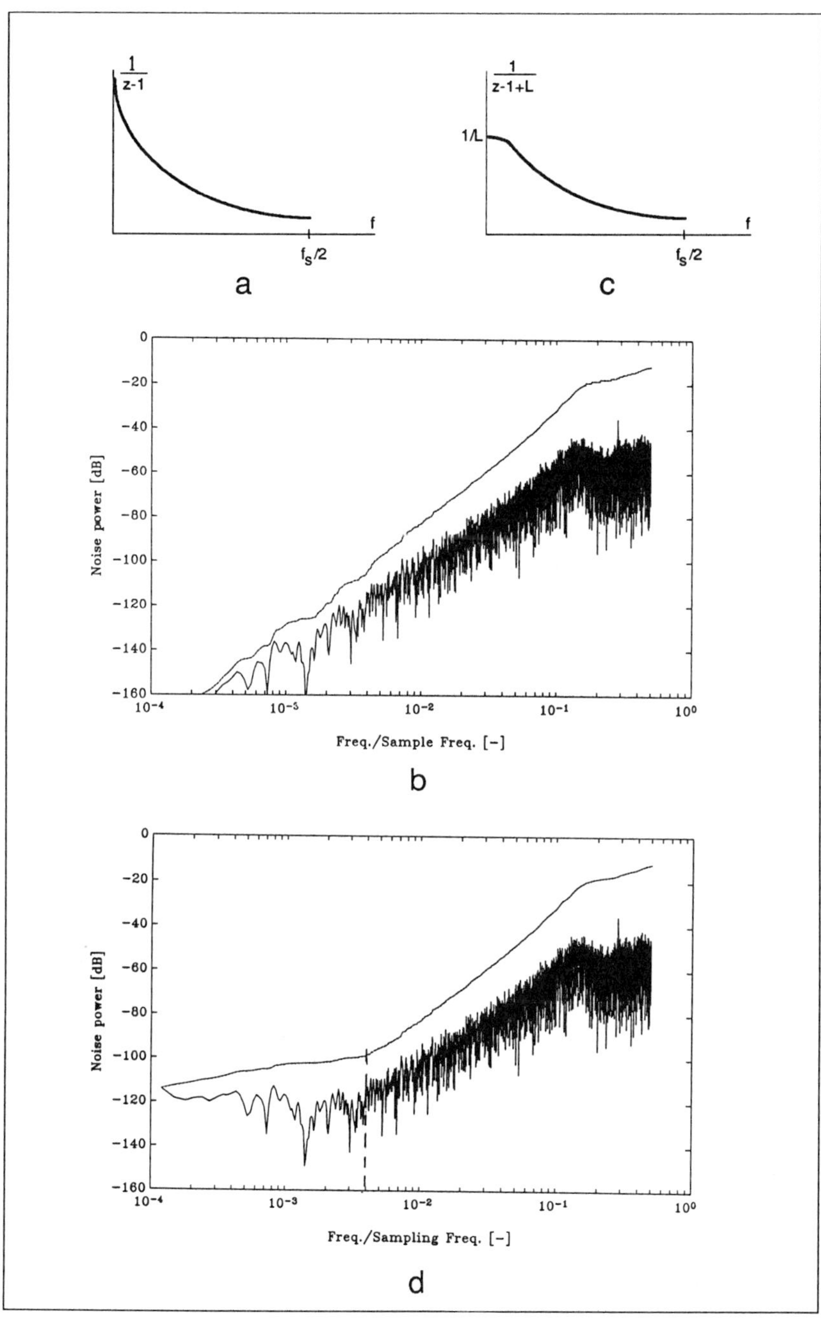

Fig. 4.15:   *a) the gain of an ideal integrator*
b) ideal quantisation noise spectrum
c) the gain of a real integrator
d) real quantisation noise spectrum

The upper curves in Figs. 4.15.b. and d. show the total integrated in-band quantisation noise power as functions of the lowpass filter frequency $f_d/2$ (see Fig. 4.1.) For small lowpass filter frequencies below the zeros expressed by (4.12), there is a large difference between these curves. But for lowpass filter frequencies well above the frequencies f_{pi}, the difference between the upper curves in Figs. 4.15.b. and d. is negigible. This implies that the increase of the in-band quantisation noise power is negligible, provided that the frequencies expressed by (4.12) are well below the lowpass filter frequency:

$$\frac{L_i}{2\pi} \cdot f_s \ll \frac{f_s}{2 \cdot OR} \quad \text{and thus:} \quad A_{DC,i} \gg OR/\pi \qquad i=1..n \qquad (4.13)$$

Although the low-frequency noise power spectrum in Fig. 4.15.d. seems much larger than in Fig. 4.15.b, the total in-band noise power is not significantly increased due to the integrator leakages, provided that each integrator in Fig. 4.1. meets condition (4.13). E.g, for an Oversampling Ratio of 100, amplifier DC gains in the order of 40 dB are required. This is not difficult to obtain in practice.

The finite integrator DC gains cause a second problem called **low signal level distortion** [14]. In section 3.4.c, Fig. 3.32, it was pointed out that input signals with a too small amplitude are not transferred to the output. The minimum signal amplitude is in the order of

$$V_{min} \approx V_{ref}/|H(f)| \qquad (4.14)$$

where f is the input signal frequency and H(f) is the filter gain (see Fig. 3.23.a).

Due to the finite integrator DC gains in Fig. 4.1, the filter gain for a DC input signal (H(f=0)) is finite. DC input signals with a magnitude smaller than $V_{ref}/H(f=0)$, are not transferred to the ADC output. This is shown in Fig. 4.16. When a quasistatic signal is applied to the ADC input, the output is as shown in Fig. 4.16.b. This results in cross-over distortion [14].

E.g. for a 14 bit ADC, a DC filter gain H(f=0) of 16384 is required to obtain a resolution below a half bin. For a Second-order Sigma-Delta modulator with two integrators, this condition is met when the DC gain of both integrator amplifiers is larger than 128. For higher-order modulators, even smaller DC gains can be tolerated.

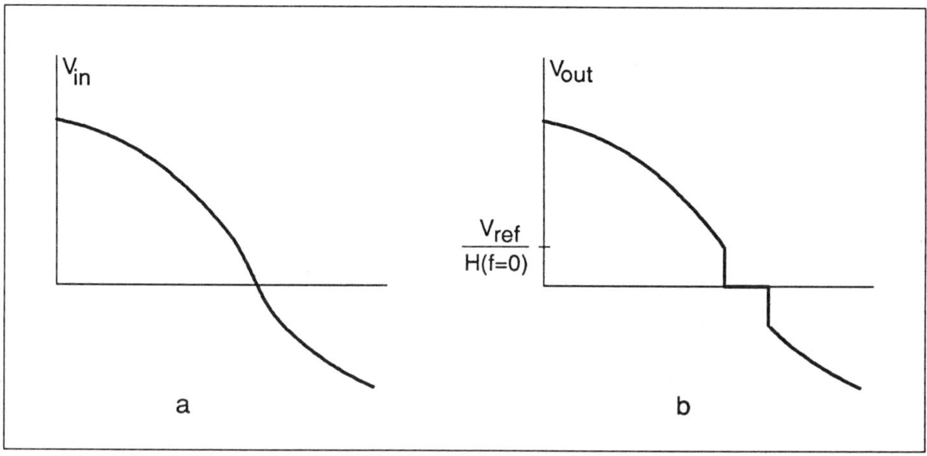

Fig. 4.16: The mechanism of "Low Signal-level Distortion"
a) modulator input signal and b) output signal

4.3.e. The signal degradation due to clock feedthrough

In Fig. 4.17.a. the schematic of a stray-insensitive Switched-Capacitor integrator [6] is depicted. Due to clock feedthrough [15-16], a fraction of the channel charge of S_1 and S_2 flows into the sampling capacitance C_1 at the falling edge of ϕ_1. In a similar way, a parasitic charge is placed on the integration capacitance C_2 at the falling edge of ϕ_2. For single nMOS or pMOS switches, these charges are in the order of 0.1 pC. For complementary switches, the clock feedthrough charges are in the order of 0.05 pC. The injected charges are nonlinear functions of the switch terminal voltages [18].

The signal degradation due to these parasitic charges can be described by an equivalent input voltage, equal to

$$v_{in,eq} = \frac{\text{sum of all the injected charges}}{C_1} \tag{4.15}$$

Note that increasing C_1 to decrease this equivalent input voltage yields no improvement: increasing C_1 requires smaller switch on-resistances and therefore larger switch transistor sizes to meet the speed requirements. The injected charges will increase as well.

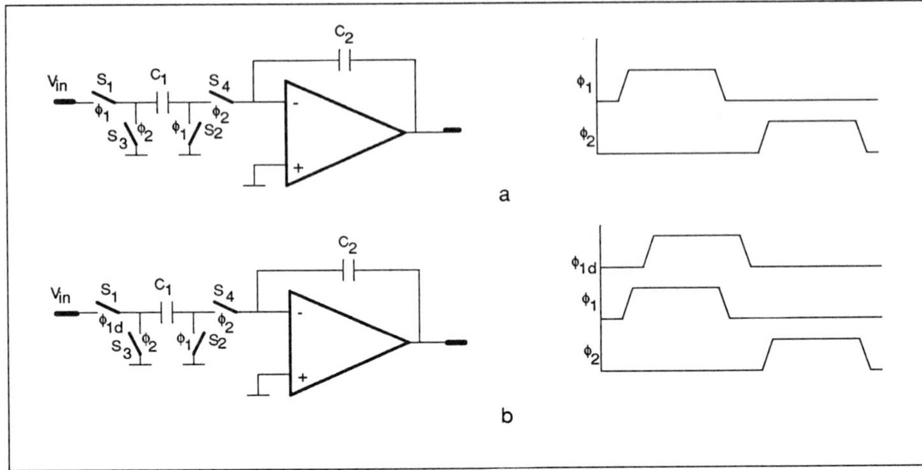

Fig. 4.17:
a) a Stray-insensitive Switched-Capacitor integrator
b) an improved version without clock feedthrough induced distortion

At the falling edge of ϕ_1, the terminal voltages of S_2 are both equal to ground. Hence, the charge injected by this switch is the same for each clock cycle. Its equivalent input voltage causes an integrator offset. In a similar way, at the falling edge of ϕ_2, the terminal voltages of S_3 and S_4 are approximately equal to ground and the charges injected by these switches are invariant as well.

For switch S_1 however, the terminal voltages equal the input voltage V_{in} at the falling edge of ϕ_1. The charge injected into C_1 by this switch is a nonlinear function of the input voltage and causes harmonic distortion.

A solution to prevent this signal degradation is shown in Fig. 4.17.b. [19]. This integrator uses a three-phase clock where ϕ_{d1} is a delayed version of ϕ_1. By opening S_1 a little later than S_2, one side of C_1 is disconnected at the moment S_1 is opened. The channel charge of S_1 has to flow into the input voltage source.

4.3.f. The signal degradation due to the settling times of the reference voltages

Consider the Switched-Capacitor integrator of Fig. 4.14.a. During ϕ_2, a charge is placed on C_1, equal to

$$Q = C_1 \cdot (V_{ref} - V_{in}) \quad \text{if } y(k)=1 \tag{4.16.a}$$

$$= C_1 \cdot (-V_{ref} - V_{in}) \quad \text{if } y(k) = -1 \tag{4.16.b}$$

Note that these charges vary with the input voltage V_{in}. Due to the finite settling times of the integrator amplifier and the reference voltage buffers or due to switch on-resistances, this charge transfer is incomplete:

$$Q = (1 - \varepsilon^+) \cdot C_1 \cdot (V_{ref} - V_{in}) \quad \text{if } y(k)=1 \tag{4.17.a}$$

$$= (1 - \varepsilon^-) \cdot C_1 \cdot (-V_{ref} - V_{in}) \quad \text{if } y(k) = -1 \tag{4.17.b}$$

In this expression, ε^+ and ε^- are the relative charge transfer errors. Due to a difference between the settling times of the reference voltage buffers or due to different on-resistances of S_1 and S_2, ε^+ can be different from ε^-.

In standard Switched-Capacitor filters, incomplete charge transfers cause no harmonic distortion, as long as the charge transfer errors (ε) are signal independent [5] [18].

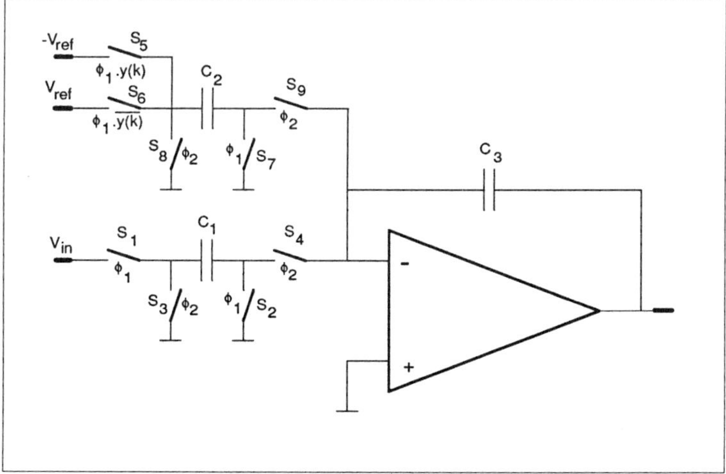

Fig.4.18: An improved Switched-Capacitor integrator and summator

In a Sigma-Delta modulator on the other hand, a difference between ε^+ and ε^- causes harmonic distortion, even when the relative charge transfer errors are constant. In appendix 4.A, an expression for this distortion is calculated:

$$HD_2 = \frac{1}{4}\cdot\frac{A_{in}}{V_{ref}}\cdot(\varepsilon^- - \varepsilon^+) \qquad (4.18)$$

where A_{in} is the ADC input signal amplitude. E.g. when ε^+ and ε^- differ with 10% of their value, the charge transfer error has to be below 0.16% to ensure 14 bit accuracy. This requires fast, accurate reference voltage output buffers.

A solution to avoid this distortion is shown in Fig. 4.18. For this circuit, the charge drawn from the reference voltages during ϕ_2 is given by:

$$Q = (1-\varepsilon^+)\cdot C_2\cdot V_{ref} \qquad \text{if } y(k)=1 \qquad (4.19.a)$$

$$= (1-\varepsilon^-)\cdot C_2\cdot(-V_{ref}) \quad \text{if } y(k)=-1 \qquad (4.19.b)$$

This charge is independent of V_{in}. With a similar calculation, it is easy to show that a difference between ε^+ and ε^- causes no harmonic distortion.

4.3.g. The signal degradation due to component noise

Besides the quantisation noise and the aperture noise, component noise (especially amplifier 1/f noise) is one of the the major signal corrupting factors. For each integrator of Fig. 4.1, the internal noise sources can be described by an equivalent noise voltage at the integrator input. Since for in-band signal frequencies the transfer function of the first integrator is large, the noise contributions from the other integrators are negligible compared with the signal degradation due to the first integrator noise.

When the integrators are implemented as shown in Fig.4.18, the equivalent input noise of each integrator is due to switch noise, due to amplifier noise and due to noise from the reference voltages.

i. The switch noise

Consider the Switched-Capacitor integrator of Fig. 4.18. The thermal noise from the switch on-resistances is sampled and aliased into the signal band [18] [21-22]. The equivalent in-band noise power spectrum at the integrator input due to the switches S_1 to S_4 (see Fig.4.18.) can be approximated by:

$$dv_{in,eq}{}^2 \approx \frac{4kT}{C_1\cdot f_s}\cdot df \qquad (4.20)$$

and the in-band noise power equals

$$N_{eq,in} = \frac{2kT}{C_1} \cdot \frac{1}{OR} \tag{4.21}$$

In the same way for the in-band noise power due to S_5 to S_9:

$$N_{eq,in} = \frac{2kT}{C_2} \cdot \frac{1}{OR} \cdot (\frac{C_2}{C_1})^2 \tag{4.22}$$

Compared with standard noise expressions for Switched-Capacitor circuits, it is striking that only a fraction 1/OR of the total switch noise is sampled into the signal band. E.g. for an accuracy of 14 bit and with an Oversampling Ratio of 100, capacitances C_1 and C_2 of about 0.3 pF are sufficient. Note that sampling capacitances of the order of 30 pF are required in standard Switched-Capacitor filters to obtain the same noise performance.

ii. The amplifier noise

The amplifier high-frequency white noise is sampled by S_4 and S_9 and aliased into the signal band. The equivalent input noise power can be obtained from standard Switched-Capacitor noise calculation techniques [18] [21-23]:

$$N_{eq,in} = (1+\frac{C_2}{C_1})^2 \cdot \frac{\text{total integrated opamp white noise power}}{OR} \tag{4.23}$$

Again, only a fraction 1/OR is aliased into the signal band. This implies that the compensation- and load capacitances of the amplifiers can be a factor OR smaller than in standard Switched-Capacitor circuits.

In most practical designs, the largest noise contribution is due to the opamp 1/f noise. Since the 1/f noise power distribution is small at high frequencies, the amount of high-frequency 1/f noise aliased into the signal band is negligible. The equivalent in-band input noise power due to the opamp 1/f noise equals:

$$N_{eq,in} = (1+\frac{C_2}{C_1}) \cdot \text{the integrated in-band amplifier 1/f noise power} \tag{4.24}$$

Note that there is no factor "1/OR" in this expression. For example when C_1 and C_2 are equal, a 14 bit accuracy requires an in-band opamp 1/f noise voltage below 25 μVolt. This is hard to obtain in a CMOS technology, especially when the speed requirements demand short-channel transistors.

For output wordlengths over 14 bit, special design techniques such as "correlated double sampling" [17] [23] or "chopper amplifiers" [24] have to be applied to further reduce the ADC 1/f noise.

iii. The noise from the voltage references

Besides the requested DC reference voltage, the output of the reference voltage generators in Fig. 4.18. contain both white and 1/f noise. This causes an equivalent in-band noise power, equal to

$$N_{eq,in} = (\frac{C_2}{C_1})^2 \cdot \frac{\text{total integrated white noise power of one reference voltage}}{OR} +$$

$$(\frac{C_2}{C_1})^2 \cdot \text{integrated in-band 1/f noise power} \qquad (4.25)$$

In the same way as for the switches and the amplifiers, only a fraction 1/OR of the white noise power is in the signal band.

4.3.h. The signal degradation due to aliasing of spurious signals

As indicated in Fig. 3.23.a, the ADC input signal passes an anti-aliasing filter first to remove the high-frequency signal contributions. When the ADC input signal is corrupted after the anti-aliasing filter by high-frequency spurious signals, some of these will be aliased into the signal band and cause signal degradation.

In mixed analog-digital integrated circuits, the supply voltages can vary over several hundreds of milivolts, mainly due to current spikes drawn by the digital circuits [26-28]. Since the low-frequency supply impedance is normally small, this supply voltage noise mainly consists of high-frequency contributions, around multiples of the digital clock frequency. Due to the limited PSRR of the opamps [26] or via the switch capacitances `[27], this supply noise can cause signals at the integrator outputs. Sampling these high-frequency contributions yields aliased in-band spurious signals that degrade the signal quality.

With separated supply lines for the analog and the digital circuits, the supply crosstalk can be significantly reduced. However, there is always a connection between the analog and digital VSS via the substrate contacts. Connecting different supply voltages to the substrate causes considerable substrate currents which influence the analog circuits.

A possible solution for this problem is shown in Fig. 4.19. The analog circuits are powered by the analog supply lines. The digital circuits use the analog supply lines for the substrate and well contacts and the digital supply lines for those connections that draw current [28]. In this way, the substrate and wells are connected only to the analog supply voltages.

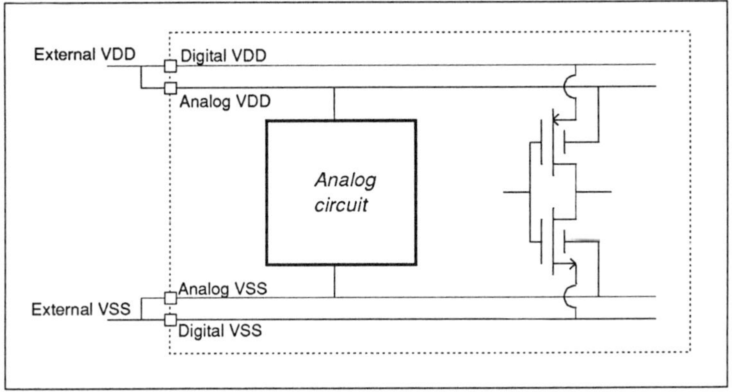

Fig. 4.19: The supply distribution schematic with modified digital gates

With differential circuits, the supply noise can be further reduced with about 20 dB [27].

4.3.i. Conclusions

In general, a Switched-Capacitor Sigma-Delta modulator is much more tolerant to analog circuit imperfections than other Switched-Capacitor circuits. Integrator gain deviations of over 20% can be tolerated. This explains why for the same amplifier GBW, the clock frequency in a Sigma-Delta modulator can be much larger than in a Switched-Capacitor filter. Sampling frequencies in the order of 10 to 20 MHz have been reported in the literature [9] [10].

Offset voltages or limited DC gains cause no major problems. For modulators with an order larger than one, integrator DC gains of 40 dB are sufficient. The distortion due to clock feedthrough can be eliminated with a three-phase clock as shown in Fig. 4.17.b. The white noise requirements are also less stringent than for a Switched-Capacitor filter.

Besides the quantisation noise and the aperture noise, the 1/f noise and the harmonic distortion due to the first integrator nonlinear gain are the major signal-degrading effects. Especially the 1/f noise is hard to reduce. Without special techniques such as Correlated Double Sampling or Chopping, the accuracy is limited by the 1/f noise to about 14 to 16 bit in a standard CMOS process.

4.4. A PRACTICAL REALISATION OF A FOURTH-ORDER ADC

In this section, a realised Switched-Capacitor Fourth-order Sigma-Delta ADC is presented. It achieves 14 bit accuracy for an output rate f_d of 500 ksample/sec. The internal clock rate f_s is 32 MHz and the oversampling ratio equals 64 [12].

4.4.a. The principle schematic

In Fig. 4.20, the basic schematic of the Fourth-order modulator is repeated. The single-bit DAC in the feedback path is split up in four DACs, one for each integrator stage.

Fig. 4.21, shows the simulated output spectrum for a sinusoidal input signal with an amplitude equal to $0.7.V_{ref}$. As can be seen, when the modulator is followed by a digital lowpass filter with a cut-off frequency $f_d/2$ of 250 kHz (see Fig. 4.1), the in-band quantisation noise power equals -96 dB [*]

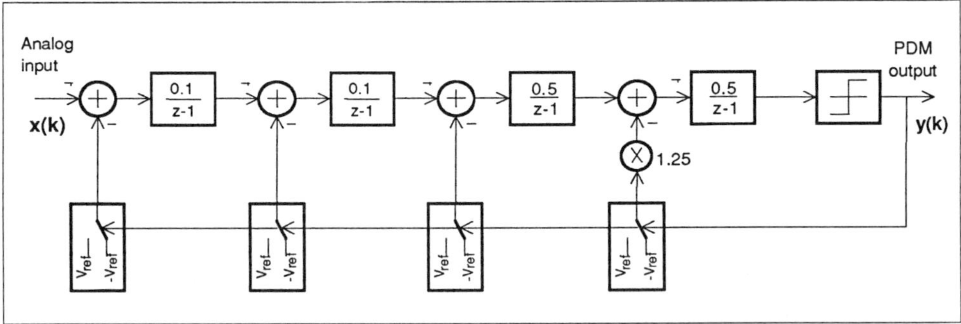

Fig. 4.20. Principle schematic of the realised ADC

---

[*] A sinusoidal voltage with an RMS value equal to V_{ref} corresponds to a power of 0 dB.

In section 4.2.c, it is pointed out that the modulator is stable for signals between -$0.75.V_{ref}$ and $+0.75.V_{ref}$. Hence, the maximum allowed signal amplitude is 0.75 V_{ref}, which corresponds with a signal power of -5.5 dB. When taking only the quantisation noise into account, the SNR for a full-scale input is 90.5 dB. Since a 14 bit ADC requires an SNR of 85.7 dB, there is a margin of about 5 dB.

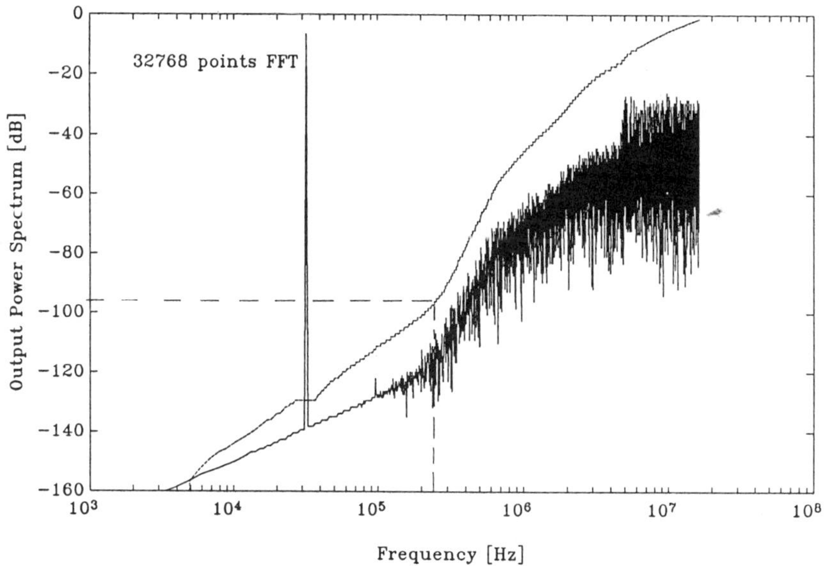

Fig. 4.21: The simulated modulator output spectrum. The clock frequency f_s is 32 MHz. The input signal is a sinususoidal voltage with a frequency of of 31.25 kHz and with an amplitude of 0.7.V_{ref}

In Fig. 4.22.a, the block schematic of the ADC is shown. One can distinguish four Switched-Capacitor integrators, a comparator, an output buffer and some clock logic.

The integrators and summators of Fig. 4.21. are realised fully differentially with Switched-Capacitor circuits. A detail of one integrator is shown in Fig. 4.22.b. The opamp, the capacitors C_1 to C_4 and the switches S_1 to S_8 form a stray-insensitive non-inverting integrator [6]. During ϕ_1, the output of the previous integrator (v_{in}) is sampled by C_1 and C_2. During ϕ_2, the charge on these capacitors is transferred to C_3 and C_4.

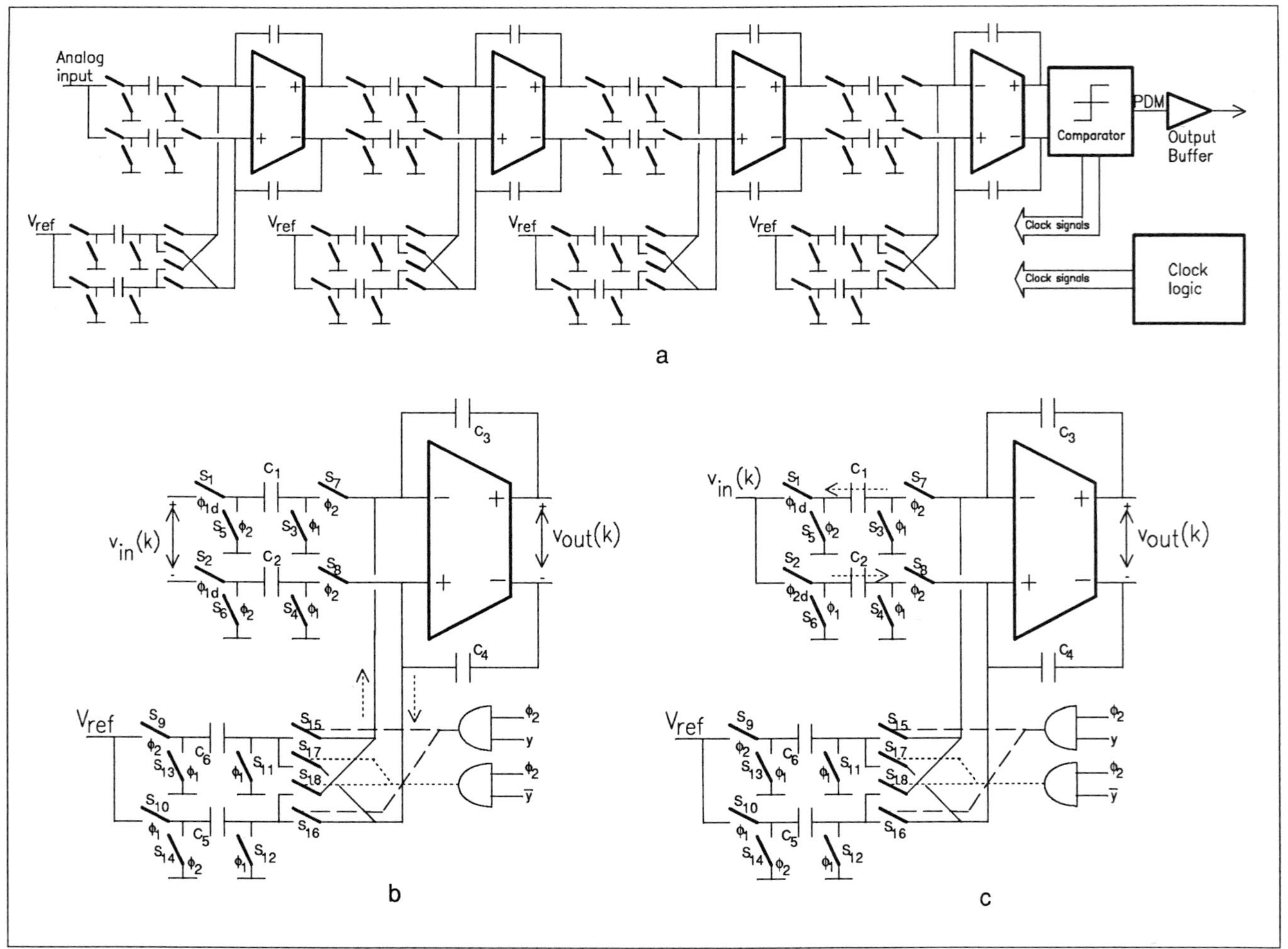

Fig. 4.22: a) A detailed schematic of Fig. 4.20
b) a detail of the last three integrators
c) a detail of the first integrator

Capacitors C_5 and C_6 form a one bit DAC with the switches S_9 to S_{18}. This can be explained as follows: assume that C_5 and C_6 are equal. During ϕ_1, C_5 is precharged to V_{ref} while C_6 is discharged. During ϕ_2, C_5 is discharged while C_6 is charged to V_{ref}. Switches S_{15} and S_{16} are controlled by a logical AND of the signals ϕ_2 and y(k). When the comparator output signal y(k) is high, these switches are closed during ϕ_2 and a charge equal to $C_5.V_{ref}$ flows in the direction of the arrows in Fig. 4.22.b. When on the other hand, y(k) is low, S_{17} and S_{18} are closed and the direction of the charges is reversed. In general, the output after ϕ_2 can be expressed as:

$$v_{out}(k+1) = v_{out}(k) + \frac{C_1}{C_3}.v_{in}(k) - 2.\frac{C_5}{C_3}.V_{ref} \quad \text{if } y(k)=1 \qquad (4.26.a)$$

$$v_{out}(k+1) = v_{out}(k) + \frac{C_1}{C_3}.v_{in}(k) + 2.\frac{C_5}{C_3}.V_{ref} \quad \text{if } y(k)=-1 \qquad (4.26.b)$$

This can be reformulated in a closed expression:

$$v_{out}(k+1) = v_{out}(k) + \frac{C_1}{C_3}.v_{in}(k) - 2.\frac{C_5}{C_3}.V_{ref}.y(k) \qquad (4.27)$$

The circuit of Fig. 4.22.b. performs the functions of a DAC, a summator and an integrator.

The switches of Fig. 4.22.b. are realised with CMOS transmission gates [17]. The switch on-resistance is 500 Ohm. In order to avoid the signal degradation due to clock feedthrough described in section 4.3.e, switches S_1 and S_2 are opened a little later than S_3 and S_4. (see Fig. 4.17.b.). For that purpose, a delayed clocksignal ϕ_{1d} is generated by the clock logic.

The three last integrators in Fig. 4.22.a. are realised according to the circuit in Fig. 4.22.b. For the first integrator, there is an additional complication: the input is a single-ended signal. This integrator is realised according to Fig. 4.22.c. Compared with the three other integrators, the switches S_2 and S_6 are controlled with different clock signals. During ϕ_1, C_1 is charged to V_{in} while C_2 is discharged. During ϕ_2, the charge stored on C_1 is transported to C_3 while C_2 is charged to V_{in}. Hence, during ϕ_2, C_1 and C_2 charge C_3 and C_4 with a charge equal to $C_1.V_{in}$, in the direction of the arrows in Fig. 4.22.c. The integrator output after ϕ_2 can be written as:

$$v_{out}(k+1) = v_{out}(k) + 2.\frac{C_1}{C_3}.V_{in} - 2.\frac{C_5}{C_3}.V_{ref}.y(k) \qquad (4.28)$$

In order to avoid distortion due to clock feedthrough, S_1 and S_2 in Fig. 4.22.c. are clocked with delayed clock signals ϕ_{1d} and ϕ_{2d}.

The capacitor values for the different integrators are summarised in Table 4.1.

integrator #	C_1-C_2	C_3-C_4	C_5-C_6
1	1 pF	10 pF	1 pF
2	1 pF	10 pF	1 pF
3	1 pF	2 pF	1 pF
4	1 pF	2 pF	1.25 pF

Table 4.1: The capacitor values for the different integrators

4.4.b: The amplifiers

In order to achieve a linear integrator gain for the first integrator, small charge transfer errors are required. Because of the large clock frequency of 32 MHz, a fast opamp with a GBW of about 100 MHz has to be used. The DC gain has to be large as well. Fig. 4.23. shows the schematic of an opamp that meets these requirements. This is a two-stage Miller compensated differential-input differential-output amplifier. With the resistors R_1 and R_2, the common-mode output voltage is measured [30]. When this common-mode signal is different from zero, this will be detected by the differential pair M_{12}-M_{13} and corrected by the common-mode feedback loop via M_{14} to M_{21}.

Because of the cascode transistors in the output stage and because of the large values of the resistors R_1 and R_2, the DC gain of the output stage is comparable to that of a two-stage amplifier. The over-all amplifier DC gain is comparable to that of a three-stage amplifier.

In Fig. 4.24, the measured amplifier open loop gain and phase are depicted, obtained from S-parameter measurements. A GBW of 128 MHz and a phase margin of 52° are measured for a load capacitance of 5 pF.

The amplifier characteristics are collected in Table 4.2.

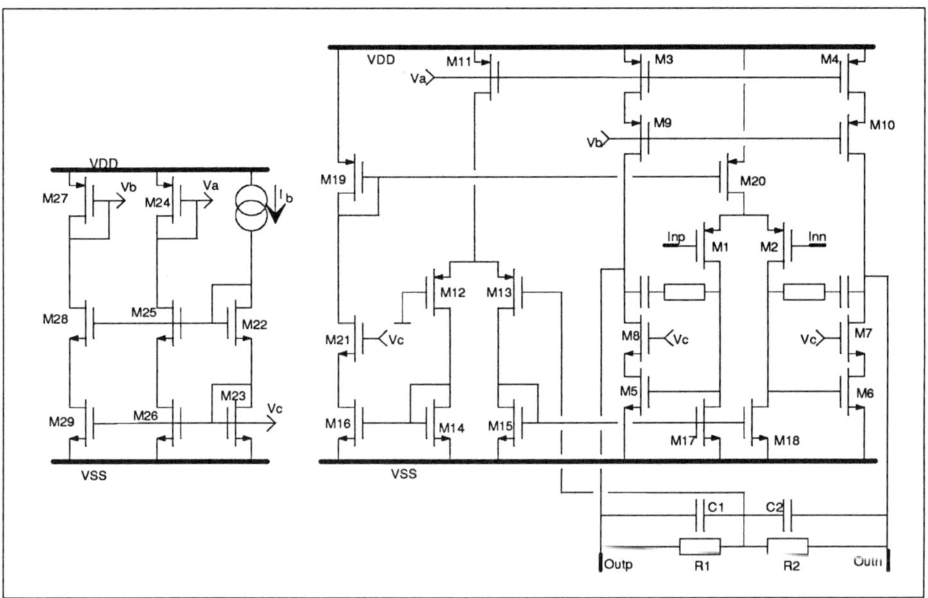

Fig. 4.23. The Two-stage Miller compensated differential-input differential-output amplifier

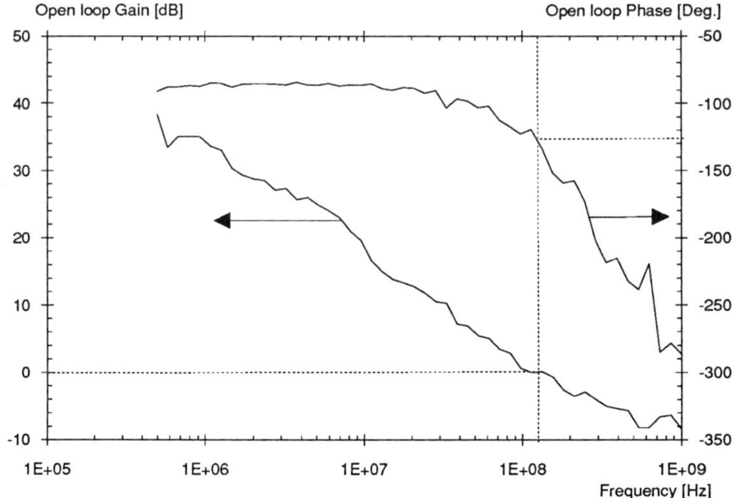

Fig. 4.24: The measured open loop amplifier gain and phase for a 5 pF load capacitance.

V_{DD}, V_{SS}	+/- 2.5 V
Load capacitance	5 pF
Input capacitance *	0.47 pF
GBW *	128 MHz
Phase Margin *	52 ^{o}
DC gain	78 dB
Supply current	6.5 mA
Supply current for the Bias circuit	1.5 mA
* obtained from S-parameter measurements	

Table 4.2: The measured amplifier characteristics

4.4.c. The comparator, the output buffer and the clock logic

Fig. 4.25.a. shows the comparator schematic. The comparator is designed according to a well-known principle with internal positive feedback [13]. During ϕ_2, the comparator input voltage is sensed by transistors M_1 and M_2. The nodes a and b are then connected to V_{DD}. During ϕ_1, switches M_8 and M_9 are closed. Transistors M_4 to M_7 form then two inverters, connected in a loop. Depending on the sign of the comparator input voltage, the positive feedback in this loop forces one of the nodes a and b to V_{SS} while the other is switched to V_{DD}. This situation is saved in a flip-flop. During the next clock phase ϕ_2, four clock signals are generated for the switches S_{15} to S_{18} in Figs. 4.22.b. and c. The 2 pF capacitances in Fig. 4.25.a. represent the capacitances of a 2 mm long metal line plus the switch gate capacitances.

The output buffer is a large CMOS inverter, designed to drive a load capacitance of 30 pF.

The clock logic of Fig. 4.25.b. consists of six inverters. From an external two-phase non-overlapping clock, the clock signals ϕ_{1d} and ϕ_{2d} plus their complements are derived. Again, the inverter loads are estimated equal to 2 pF.

Note that all the digital gates are designed with the separated clock lines as illustrated in Fig. 4.19.

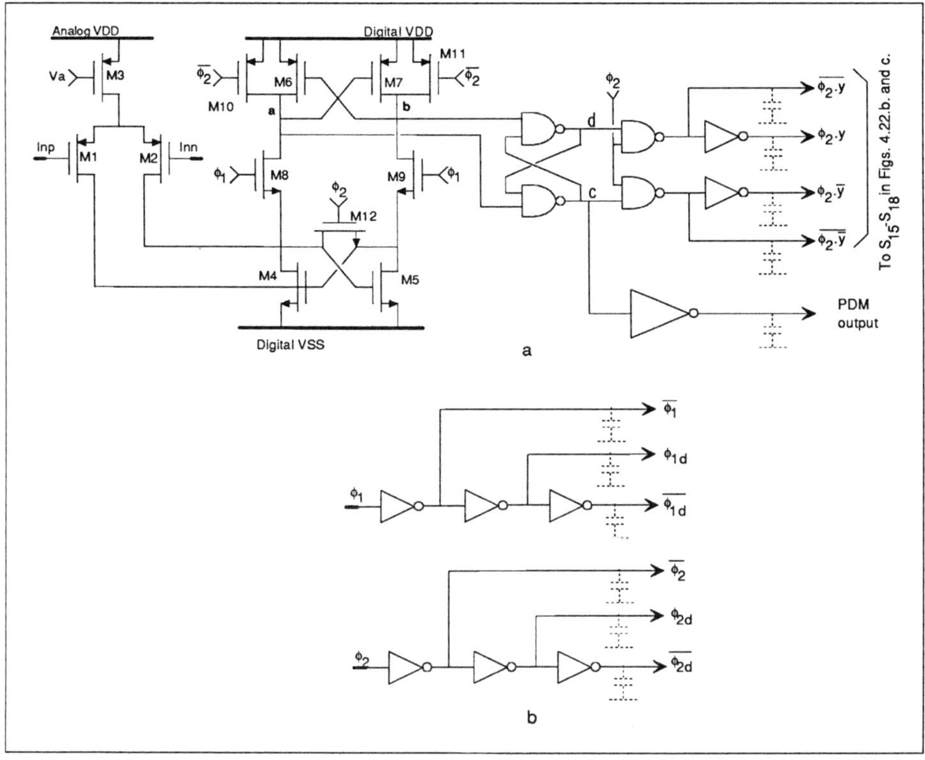

Fig. 4.25: a) The comparator and the output buffer
  b) The clock logic

Fig. 4.26. shows a timing diagram for the digital circuits. As can be seen, the clocks ϕ_{1d} and ϕ_{2d} switch about 1.5 ns after the clocks ϕ_1 and ϕ_2. About 4 nsec. after the leading edge of ϕ_1, the flip-flop output nodes c and d are switched to their final levels. 1.5 nsec. after the leading edge of ϕ_2, the four clock signals for the switches S_{15} to S_{18} are available.

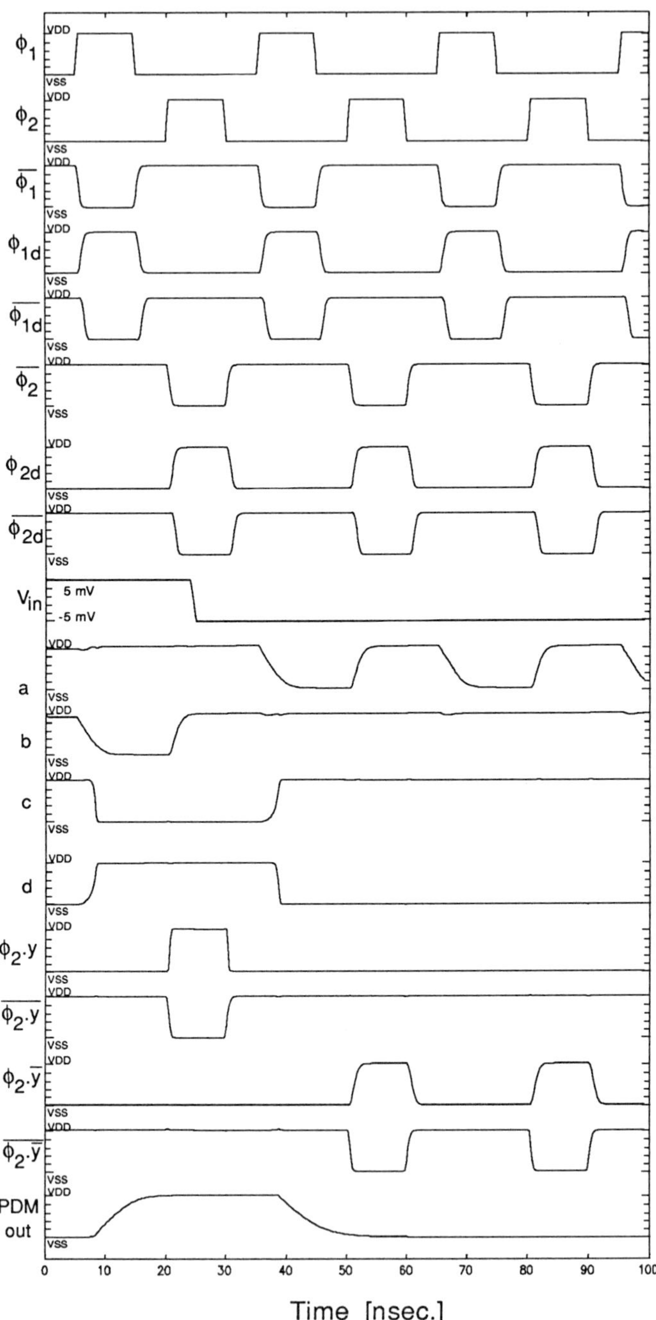

*Fig. 4.26: a simulated timing diagram of the comparator
and of the clock signals. The clock frequency is 32 MHz.
The comparator input voltage is +5 mV or -5 mV.*

4.4.d. The global realisation and the measurement results

The circuit of Fig. 4.22.a. is realised in a 1.5 μm double poly double metal CMOS process. In Fig. 4.27, a microphotograph of the chip is depicted. One can clearly distinguish the different building blocks. The die size is about 2.3 mm^2, bonding pads not included.

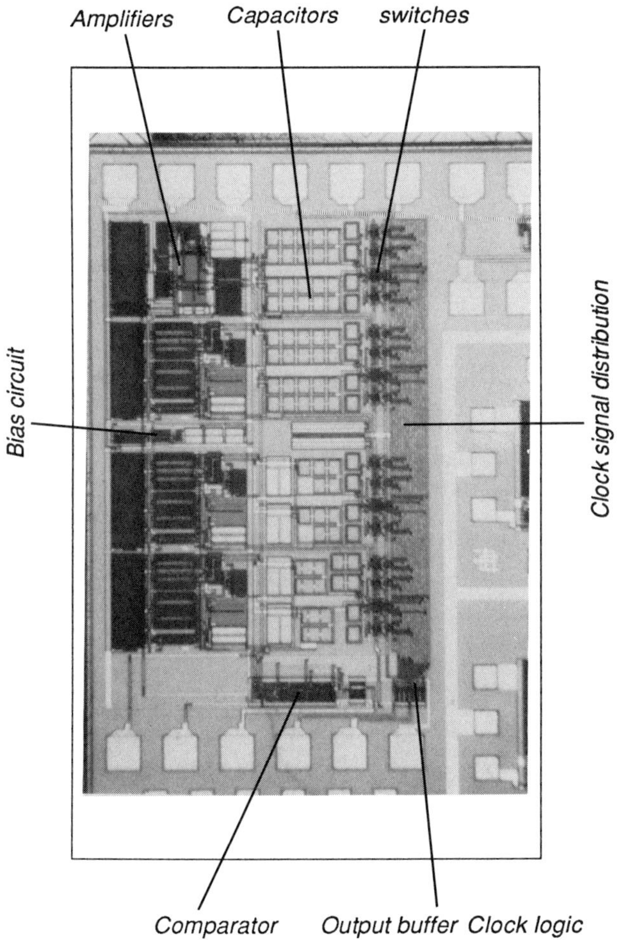

Fig. 4.27: A microphotograph of the Fourth-order ADC

The performance of this ADC is measured with a setup according to Fig. 3.45.b. Fig. 4.28.a. shows the measured output power spectrum for a clock frequency of 32 MHz, an input frequency of 40 kHz and an input amplitude of 0.6 Volt. (The reference voltage equals one Volt.) The large peaks close to DC are due to the spectral leakage from the input signal DC component (this DC component is due to the input referred offset voltage of the ADC). These lines are not generated by the ADC but by the measurement equipement.

The spectral lines around 40 kHz are from the sinusoidal input signal.

In Fig. 4.28.b, the ADC output spectrum is shown after a decimation with a factor 64.

The ADC characteristics are collected in Table 4.3.

Supply voltage	+/- 2.5 V
Supply current	31.7 mA
Reference voltage	1 V
Die Size	2.3 mm^2
Clock frequency	32 MHz
Oversampling Ratio	64
Signal bandwidth	250 kHz
	(500ks/s)
Offset voltage	30 mV
Maximum signal power *	-5.5 dB
HD_2	93 dB
HD_3	97 dB
S/(N+THD)	83 dB
Resolution	13.5 bit
* 0 dB = 1 Volt RMS	

Table 4.3: Measured Fourth-order ADC Characteristics

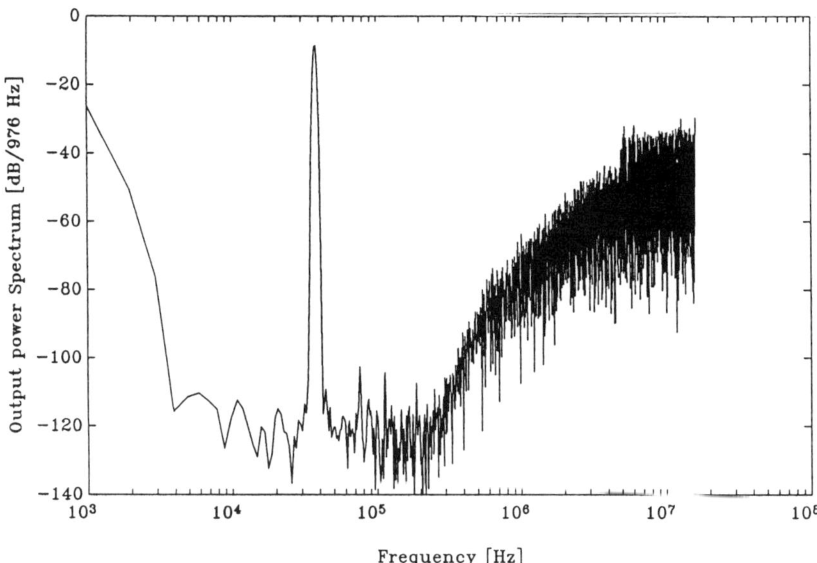

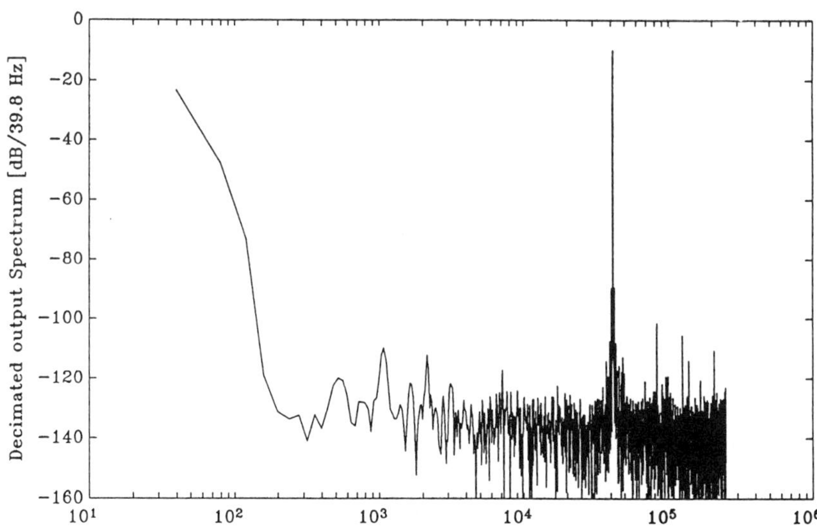

Fig. 4.28. Measured ADC output spectrum
a) before decimation
b) 64 times decimated

4.5. ALTERNATIVE APPROACHES FOR HIGHER-ORDER SIGMA-DELTA MODULATORS

The filter coefficients a_1 to a_3 of the Fourth-order modulator discussed in Section 4.2. are optimised towards a maximum stable input range R. This resulted in a loop filter with a gain, given by:

$$H(z) = \frac{1}{z-1} + \frac{a_3}{(z-1)^2} + \frac{a_2 \cdot a_3}{(z-1)^3} + \frac{a_1 \cdot a_2 \cdot a_3}{(z-1)^4} \qquad (4.29)$$

$$= \frac{(z-1)^3 + a_3 \cdot (z-1)^2 + a_2 \cdot a_3 \cdot (z-1) + a_1 \cdot a_2 \cdot a_3}{(z-1)^4} \qquad (4.30)$$

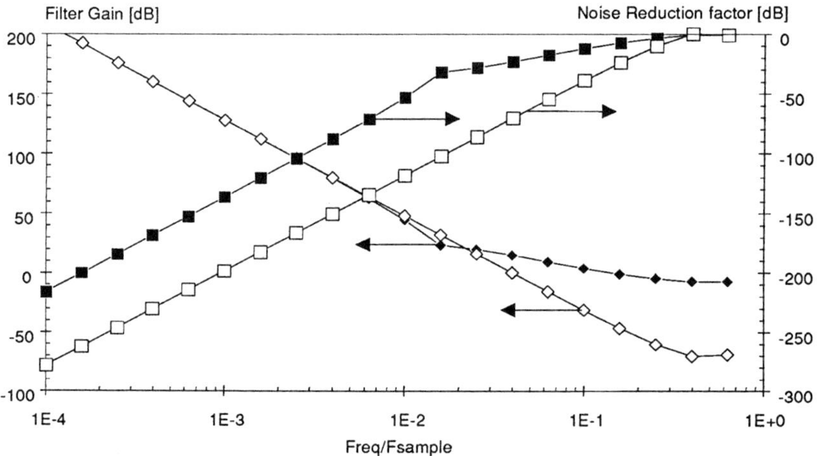

Fig. 4.29: Loop filter transfer characteristic and Quantisation noise suppression factor for an ideal Fourth-order modulator (white) and for the modulator of Fig. 4.12.a (black curves)

This filter characteristic has four poles at DC (z=1) and three zeros. Due to these zeros, the filter becomes less steep at high frequencies. In Fig. 4.29, the filter characteristic is compared with the ideal filter characteristic, without zeros. As can be seen, the real filter is less steep at high frequencies.The noise reduction factors of the real filter and of the ideal filters are also depicted in Fig. 4.29. Due to the filter zeros, the quantisation noise suppression is about 50 dB less than for the ideal case. The smaller the coefficients a_1 to a_3 are, the lower the zero frequencies and the larger the difference is between the white and the black curves. This reduces the modulator SNR.

The position of the filter zeros, is a new constraint that determines the modultors SNR. This second constraint is in conflict with the requirement of a large stable input range: when the coefficients a_1 to a_3 are large, the stable input range is zero according to Fig. 4.8. And when these coefficients are too small, the noise suppression by the filter is less effective. The optimisation of the modulator SNR is therefore a trade off between a large input signal and a small quantisation noise.

In this section, several approaches are discussed to improve or to eliminate this trade off.

4.5.a. Higher-order Sigma-Delta modulators with other filter structures

The loop filter of Fig. 4.6. has four poles at DC. Since only the position of the zeros can be optimised, this filter has only three free design variables.

In the literature, Sigma-Delta modulators with more complex filter structures are described [4] [31]. These filters have more free design variables, allowing to optimise more variables towards an optimum compromise between a large stable input range and a large SNR.

For instance, the circuit of Fig. 4.30. [4] incorporates a "Follow the Leader" loop filter with eight filter coefficients. Since the filter gain factor is of no importance, only seven coeffcients can be optimised towards the desired modulator characteristics.

Another approach is depicted in Fig. 4.31. [31]. This is a Fifth-order modulator, with five filter poles and four zeros. By using two local feedback branches (f_1 and f_2 in Fig. 4.31), two complex pole pairs are created. The fifth pole is situated at DC. The resulting filter has eight relevant free design parameters which can be optimised towards a maximum stable input range or an optimum SNR. A Fifth-order modulator with 18 bit resolution for an Oversampling Ratio of 64 has been reported, using this technique [31].

In general, more complex filter structures allow more free design variables to be optimised. This optimisation requires extensive computer simulations, similar to the simulations discussed in section 4.2.

Sigma-Delta modulators incorporating these filters still show instabilities for non-zero state variable initial conditions. Therefore, a supervision circuit as shown in Fig. 4.11. or a proper scaling of the state variables is required.

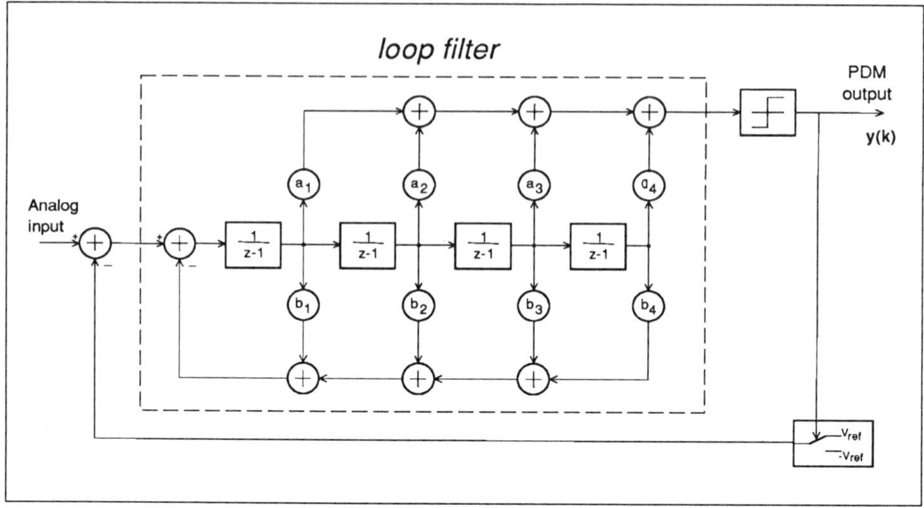

Fig. 4.30: A Sigma-Delta modulator with a "Follow the Leader" filter architecture

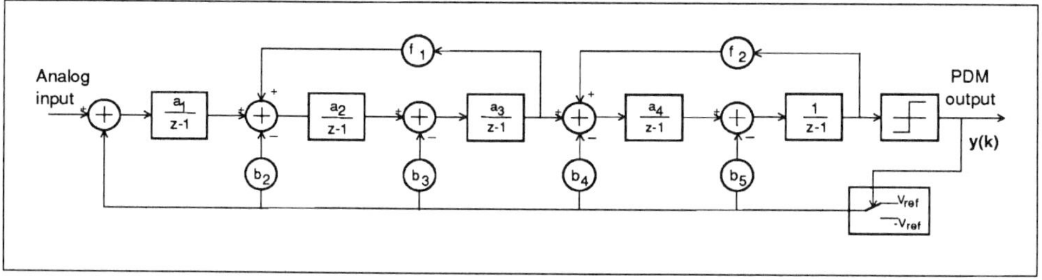

Fig. 4.31: A Sigma-Delta modulator with complex filter poles

4.5.b. Multi-bit Sigma-Delta modulator architectures

According to Fig. 3.24, a multi-bit higher-order Sigma-Delta modulator can be made unconditionally stable, provided that its input signal amplitude is reduced. In [32], it is shown that the word length of the quantiser and the feedback DAC must be at least equal to the modulator order. The required linearity of this multi-bit DAC in the feedback path is the major difficulty for this approach (see also section 3.3.c).

Fig. 4.32. shows a potential solution to overcome the nonlinearity problems of the feedback DAC [33]. When the DAC is nonlinear, the overall Sigma-Delta modulator shows an inverse nonlinearity. When now this nonlinearity is corrected with a correction curve, stored in a RAM, the resulting curve is linear. For a B-bit wide quantiser output word, the correction curve has 2^B points and a RAM with 2^B addresses is required. The correction curve can be determined at power-up by using a second reference Sigma-Delta modulator (e.g. a slow First-order Sigma-Delta modulator).

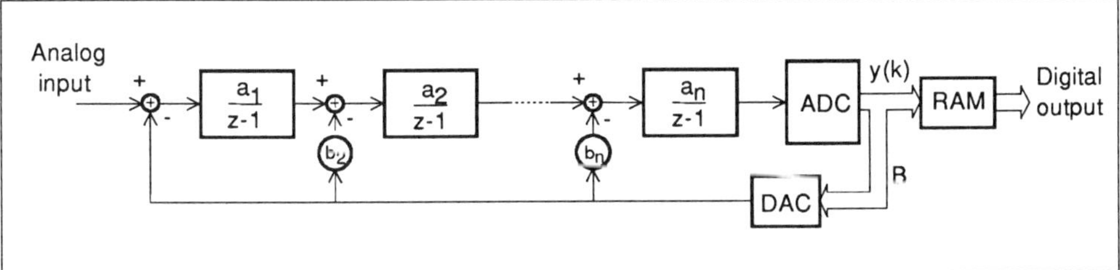

Fig. 4.32: A multi-bit Sigma-Delta ADC with digital correction of the DAC nonlinearity

Another approach uses a randomization of the DAC nonlinearities [34-35]. Consider for instance the Current-steering DAC of Fig. 4.33.a. In this 3-bit DAC, the output voltage is formed by applying a number of unity currents over an output resistor R_L. The number of unity currents is determined by the digital input word.

In a standard application, the control for the current source I_1 is determined by the LSB of the DAC input word, the switches for I_2 and I_3 are controlled by the second input bit and the switches for I_4 to I_7 are controlled simultaneously by the DAC input MSB. In other words, each switch is controlled by only bit of the input word. Assume now that there is a mismatch between the current sources and that, for instance, I_5 is too large. The DAC transfer characteristic is then as depicted in Fig. 4.33.b: for those input words where the MSB equals one, I_5 is active and the output voltage will be too large. E.g. when the input word equals 5 (101 in binary notation), currents I_1, I_4, I_5, I_6 and I_7 are selected and the output voltage is too large.

When this DAC is applied in the feedback path of a multi-bit Sigma-Delta ADC, the DAC nonlinearity will cause a nonlinear transfer characteristic of the overall ADC.

This nonlinearity can be overcome by selecting the currents in a random way: for instance, when the digital input equals 5 (101), five currents have to be switched to the output. When at each clock cycle, a random set of five currents out of the seven available current sources is connected to the output, I_5 is not always in the selected group. The mismatch of I_5 is translated into a random output voltage, thus into white noise. Although the DAC in the feedback path is nonlinear, the overall Sigma-Delta modulator characteristic is linear.

A similar approach can be used in a multi-bit Switched-capacitor DAC [34-35], where the required number of unity-capacitors can be selected in a random way, out of an available capacitor bank.

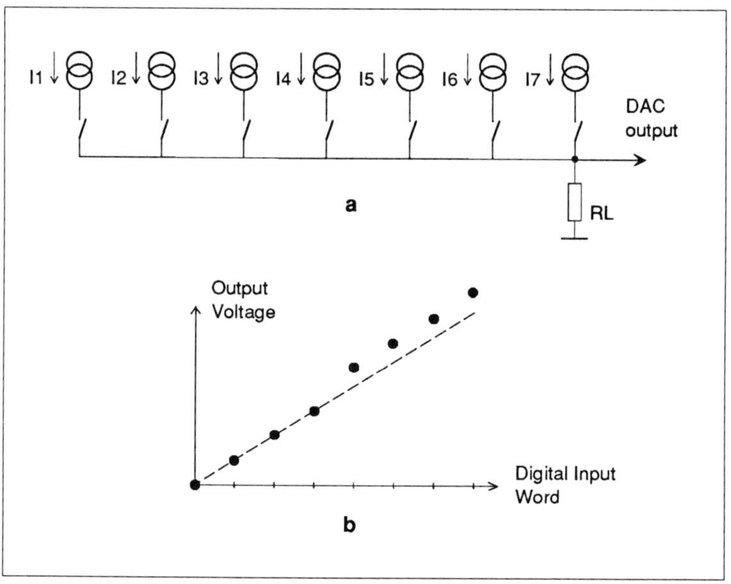

Fig. 4.33: a) a Current-steering DAC and b) the DC transfer characteristic when I_5 is too large

4.5.c. The MASH technique

The **MASH** or **multistage** technique [36-37] offers the most powerful solution to overcome the stability problems in higher-order Sigma-Delta modulators, without compromising the stable input range.

To explain this MASH principle, consider the First-order Sigma-Delta modulator in Fig. 4.34.a. The nonlinear quantiser can be modelled as a linear gain block and a source of additive quantisation noise:

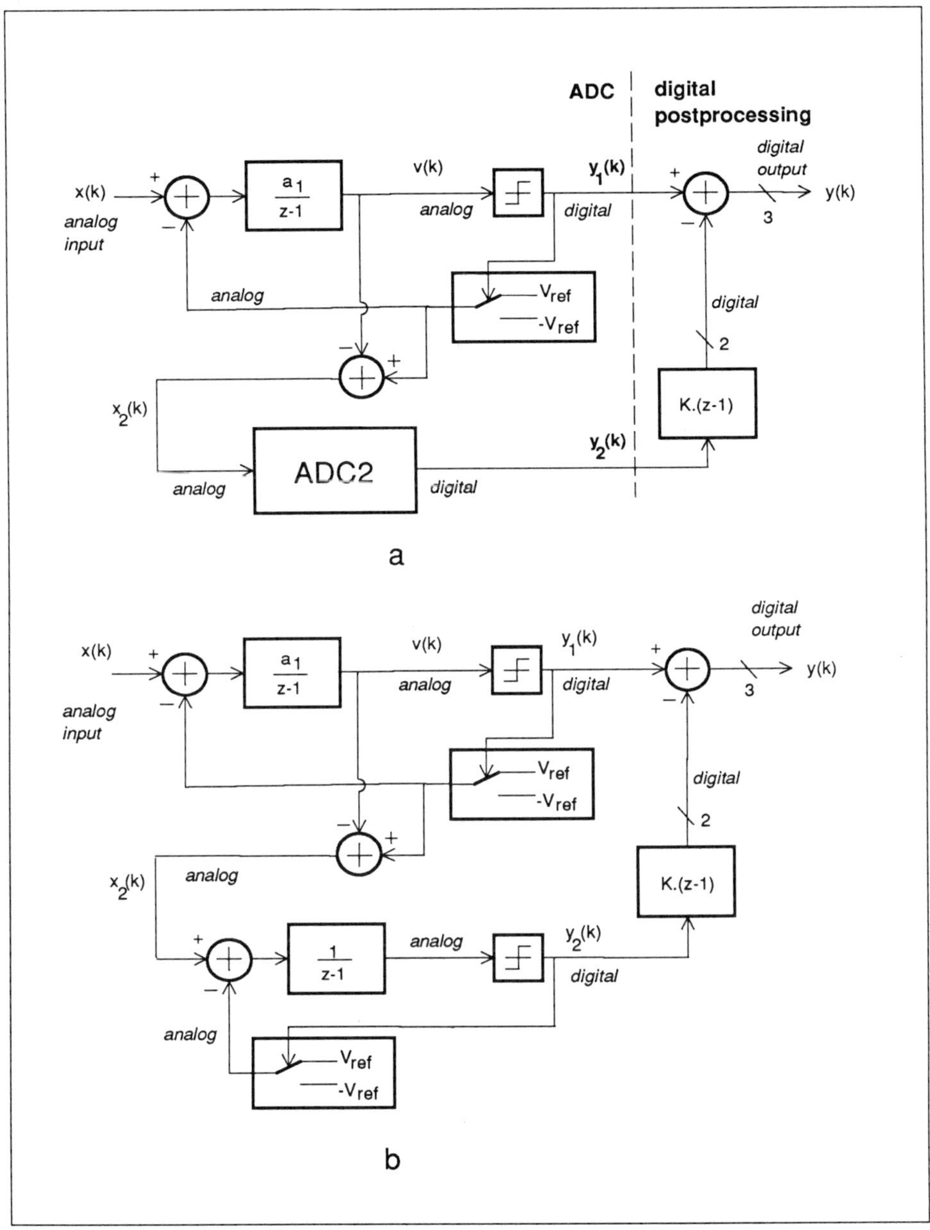

*Fig. 4.34: a) the principle of a MASH converter and b) a
practical implementation*

$$y_1(k) = v(k)/V_{ref} + e_1(k) \qquad (4.31.a)$$

or, with an expression in the frequency domain:

$$Y_1(z) = V(z)/V_{ref} + E_1(z) \qquad (4.31.b)$$

$Y_1(z)$, $V(z)$ and $E_1(z)$ are the z-transforms of $y_1(k)$, $v(k)$ and $e_1(k)$ respectively. The output $y_1(k)$ of this Sigma-Delta modulator can then be expressed as:

$$Y_1(z) = \frac{X(z)/V_{ref}}{1 + (z-1)/a_1} + \frac{(z-1)/a_1}{1 + (z-1)/a_1}.E_1(z) \qquad (4.32)$$

This PDM output signal consists of a term proportional with the input voltage x(k) and an undesired term due to the quantisation error $e_1(k)$. If the signal $e_1(k)$ would have been exactly known, it would have been easy to perform a digital error correction on $y_1(k)$, to cancel the second term in (4.32).

It is possible to extract the quantisation error $e_1(k)$ from the signals $y_1(k)$ and v(k): when v(k) is subtracted from the ADC output $y_1(k)$ as shown in Fig. 4.34.a, the result equals:

$$x_2(k) = V_{ref}.y_1(k) - v(k) = V_{ref}.e_1(k) \qquad (4.33)$$

When this voltage $x_2(k)$ is applied to a second ADC, a digital output signal $y_2(k)$ is obtained, equal to the quantisation error $e_1(k)$. With digital postprocessing, this error signal can be used to cancel the quantisation error in expression (4.32). In theory, a digital output signal can be obtained that contains no quantisation noise due to the Sigma-Delta modulator.

However, in reality, the second ADC generates quantisation noise too. More in detail, the output of the second ADC is given by:

$$y_2(k) = x_2(k)/V_{ref} + q(k) \qquad (4.34.a)$$

or, in the frequency domain:

$$Y_2(z) = X_2(z)/V_{ref} + Q(z) \qquad (4.34.b)$$

In these expressions, q(k) is the quantisation noise from the second ADC and Q(z) is its z-transform.

Since the quantisation noise from the first ADC appears at the output, after a multiplication with a factor "(z-1)", it makes sense to correct this noise error with the digital correction circuit of Fig. 4.34.a:

$$Y(z) = Y_1(z) - K.(z-1).Y_2(z) \quad (*) \tag{4.35.a}$$

With expressions (4.32) to (4.35), it is easy to see that this results in the following output signal:

$$Y(z) = \frac{X(z)/V_{ref}}{1 + (z-1)/a_1} + (1/a_1 - K).\frac{(z-1)}{1 + (z-1)/a_1}.E_1(z) -$$

$$K.\frac{(z-1)^2/a_1}{1 + (z-1)/a_1}.E_1(z) - K.(z-1).Q(z) \tag{4.36}$$

When the digital summing factor K equals $1/a_1$, the second term in this expression vanishes. The third term contains a factor "$(z-1)^2$", similar to the noise shaping factor in the quantisation noise expression of a **Second-order** Sigma-Delta modulator. Therefore, the remaining quantisation noise is shaped as the noise of a Second-order Sigma-Delta modulator. For frequencies in the signal band, this noise contribution can be reduced by increasing the Oversampling Ratio. The fourth term in (4.36) represents the noise due to the second ADC. Thanks to the digital postprocessing circuits, this quantisation noise is shaped as the noise of a First-order Sigma-Delta modulator.

This principle is known as the **MASH technique** [36] or **Multistage technique** and the circuit of Fig. 3.34.a is a **Multistage** or **Cascaded** Sigma-Delta modulator.

When the integrator gain a_1 equals one, the digital summing factor K is also equal to one. The digital postprocessing can then be realised with two delay elements and one full adder.

Consider now the interesting case where the second ADC is also a First-order Sigma-Delta modulator [36]. This situation is depicted in Fig. 4.34.b. The signals $y_1(k)$

(*) This z-domain expression (4.35.a) correspond to the following time-domain equation:

$$y(k) = y_1(k) - K.(y_2(k+1) - y_2(k)) \tag{4.35.b}$$

In theory, this signal operation cannot be realised because at the k-th clock pulse, $y_2(k+1)$ is not yet known. However, the following signal operation yields a similar result:

$$y(k) = y_1(k-1) - K.(y_2(k) - y_2(k-1)) \tag{4.35.c}$$

The signal y(k) is the same, except for one additional delay.

and $x_2(k)$ are still given by expressions (4.32) and (4.33). The second stage output $Y_2(z)$ equals:

$$Y_2(z) = \frac{X_2(z)/V_{ref} + E_2(z).(z-1)}{1 + (z-1)} \qquad (4.37)$$

and the output y(k) after digital postprocessing becomes:

$$Y(z) = Y_1(z) - K.(z-1).Y_2(z) \qquad (4.38)$$

$$= \frac{X(z)/V_{ref}}{[1 + (z-1)/a_1]} + E_1(z).\frac{(z-1).\mathbf{(1/a_1-K)} + E_1(z).(z-1)^2/a_1}{[1 + (z-1)/a_1] \ . \ [1 + (z-1)]}$$

$$- \frac{E_2(z).K.(z-1)^2}{[1 + (z-1)]} \qquad (4.39)$$

For low-frequency signals in the signal band, the denominators of this expression are approximately equal to one. When the digital summing factor K is equal to $1/a_1$, the bold term vanishes and (4.39) reduces to:

$$Y(z) = X(k)/V_{ref} + E_1(z).(1-K).(z-1)^2/a_1 + E_2(z).K.(z-1)^2 \qquad (4.40)$$

Both $E_1(z)$ and $E_2(z)$ appear at the output, after a multiplication with a factor $(z-1)^2$. Hence, for in-band frequencies, these quantisation noise terms are suppressed as in a Second-order modulator.

This conclusion is of particular interest: **a combination of two First-order modulators can have a similar quantisation noise spectrum as one Second-order Sigma-Delta modulator.**

We can extend the technique towards a cascade of three First-order modulators, as depicted in Fig. 4.35. [37] [38] [40]. The three PDM output signals $y_1(k)$ to $y_3(k)$ are digitally combined according to the following expression

$$y(k) = y_1(k) + K_1.[y_2(k)-y_2(k+1)] + K_2.[y_3(k)-2y_3(k+1)+y_3(k+2)] \qquad (4.41.a)$$

or, in the frequency domain:

$$Y(z) = Y_1(z) + K_1.(z-1).Y_2(z) + K_2.(z-1)^2.Y_3(z) \qquad (4.41.b)$$

It is easy to calculate that for frequencies in the signal band, the output Y(z) is given by:

$$Y(z) = \frac{X(z)}{V_{ref}} + K_2 . (z-1)^3 . \{ \ E_2(z) . (1-1/a_1) \ + \ E_3(z) \ \}$$

provided that $K_1 = 1/a_1 = 1/a_2$ and $K_2 = 1/a_1^2$ (4.42)

The quantisation noise $E_1(z)$ is completely cancelled. The noise contributions due to $E_2(z)$ and $E_3(z)$ are multiplied with a factor $(z-1)^3$. These noise contributions are suppressed with a similar factor as the quantisation noise of a Third-order modulator. When the integrator gains a_1 and a_2 are equal to unity, the noise contribution due to $E_2(z)$ vanishes too:

$$Y(z) = X(z)/V_{ref} + E_3(z) . (z-1)^3$$ (4.43)

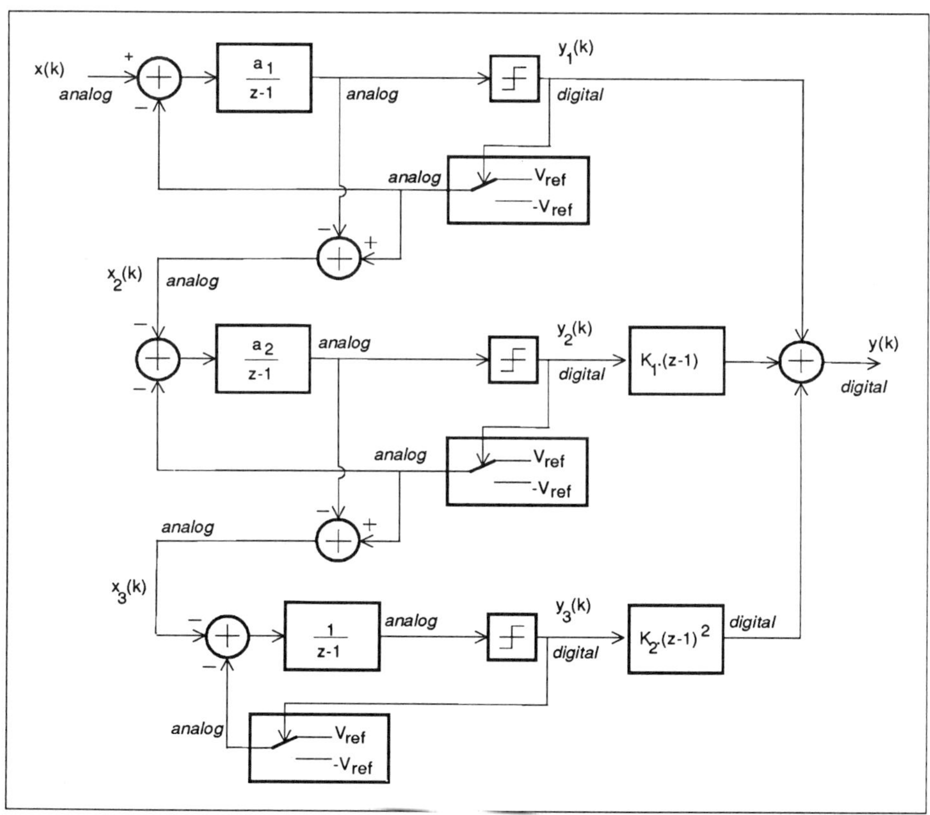

Fig. 4.35: a Three-stage MASH converter

Hence, a cascade of three First-order Sigma-Delta modulators behaves like a Third-order modulator.

It is easy to verify that the voltages at the second and third stage inputs in Fig. 4.35. are between $-V_{ref}$ and $+V_{ref}$, provided that the integrator gains a_1 and a_2 are less or equal than one. Under this condition, the Third-order Sigma-Delta modulator of Fig. 4.35. is **unconditionally stable**. The stable input range is between $-V_{ref}$ and $+V_{ref}$ and there are no low-frequency filter zeros than can reduce the SNR.

In theory, this MASH principle can be extended without limit to Sigma-Delta modulators with orders higher than three. In practice however, there is a limit due to component mismatches. To explain this, consider again the circuit of Fig. 3.34.b. This circuit behaves like a Second-order modulator, provided that the digital summing factor K equals $1/a_1$. This summing factor is determined by digital circuits, and is therefore exactly known. The integrator gain a_1 on the other hand is determined by a ratio of capacitors and is influenced by capacitor mismatches. Also when the charge transfer in this integrator is incomplete due to a finite amplifier DC gain or a finite GBW, the integrator gain a_1 deviates from its ideal value, resulting in an imperfect cancelling of the first Sigma-Delta modulator quantisation noise. This yields an additional noise term with a spectral density equal to:

$$\varepsilon . E_1(z) . (z-1)/a_1 \tag{4.44}$$

where ε is the deviation of a_1 from its ideal value. From expression (3.27), it can be seen that this additional output noise term is negligible compared with the noise given by (4.40), when

$$\varepsilon < 2K.\sqrt{3/5}. \ \pi/OR \tag{4.45}$$

For instance, for an Oversampling Ratio of 100 and K equal to one, ε can be as large 5%. This is not difficult to obtain is a modern analog silicon process.

For the Third-order modulator of Fig. 4.35. on the other hand, the additional noise due to mismatches is negligible compared with the noise given by expression (4.43), provided that

$$\varepsilon < K_2.\sqrt{3/7}. \ (\pi/OR)^2 \tag{4.46}$$

E.g. for an Oversampling Ratio of 100 and for K_2 equal to one, ε should be below 0.06%. This is impossible to realise in practice.

An alternative for this problem is shown in Fig. 4.36. [39-40] Here, the first stage is a Second-order Sigma-Delta modulator. For in-band frequencies, the output signal y(k) is given by:

$$Y(z) = Y_1(z) + K_2.(z-1)^2.Y_2(z) \qquad \text{and thus} \tag{4.47}$$

$$Y(z) = X(z)/V_{ref}.\{ 1 + K_1.K_2.(z-1)^2 \} +$$
$$E_1(z).\{ (z-1)^2.[\mathbf{1/a_1a_2-K_1K_2}] + (z-1)^3.[1/a_1a_2-K_1K_2b_2/a_1] \}$$
$$+ E_2(z).K_2.(z-1)^3 \tag{4.48}$$

$E_1(z)$ and $E_2(z)$ are the quantisation noise z-transforms of the first and second stage, respectively. When the product of the gain factors $K_1.K_2.a_1.a_2$ equals one, the bold factor vanishes and (4.48) reduces to:

$$Y(z) = X(z)/V_{ref}.\{ 1 + K_1.K_2.(z-1)^2 \} +$$
$$E_1(z).(z-1)^3.K_1K_2[1-b_2/a_1] + E_2(z).K_2.(z-1)^3. \tag{4.49}$$

Both $E_1(z)$ and $E_2(z)$ are multiplied with a factor $(z-1)^3$ in this expression. This cascade of a Second- and a First-order modulator therefore behaves like a Third-order Sigma-Delta modulator.

Note that expression (4.48) contains no noise terms multiplied with $(z-1)^1$ that have to be cancelled.

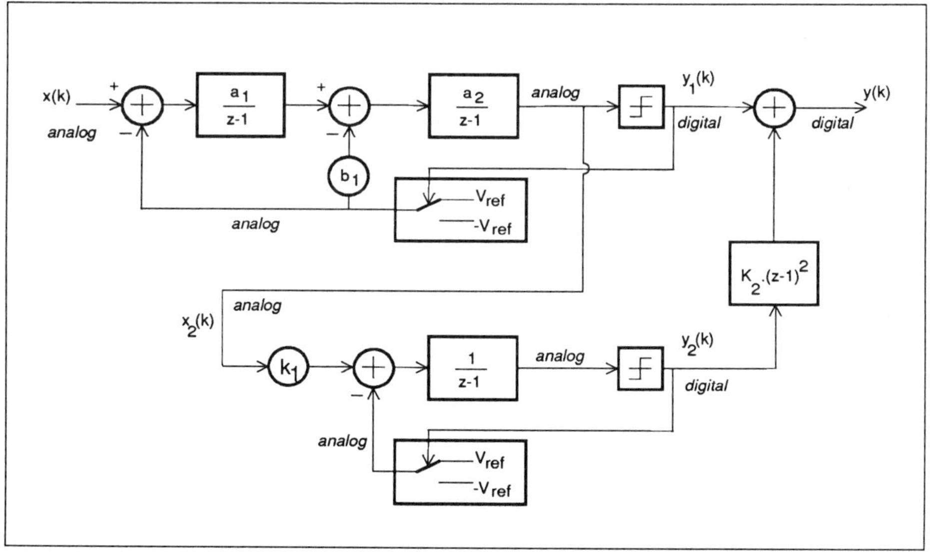

Fig. 4.36: A Two-stage Third-order MASH converter

The factors a_1, a_2 and K_1 are determined by capacitor ratios. Capacitor mismatches or incomplete charge transfers will cause an additional output noise voltage, with a spectrum given by:

$$\varepsilon.K_2.E_1(z).(z-1)^2 \tag{4.50}$$

In contrast with the spectrum given by (4.44), the output noise due to mismatches is already shaped like the quantisation noise of a Second-order Sigma-Delta modulator. A capacitor mismatch with a factor ε causes no significant SNR degradation for this Third-order modulator, as long as

$$\varepsilon < \sqrt{5/7}.(\pi/OR) \tag{4.51}$$

4.5.d. A Multi-bit MASH Converter

The A-to-D converter of Fig. 4.36. still contains one problem: when the product of the integrator gains a_1 and a_2 equals one, the signal amplitude at the second integrator output is as depicted in Fig. 3.27.c. For input signal values $x(k)$ close to V_{ref} or $-V_{ref}$, the signal amplitude of $x_2(k)$ is much larger than V_{ref} and exceeds the stable input range of the second stage ADC. Therefore, it is necessary to scale the signal $x_2(k)$ by reducing the integrator gains a_1 or a_2. A further reduction of the signal amplitude of $x_2(k)$ is achieved when the gain factor K_1 is less than one.

The cancellation of the noise due to $e_1(k)$ requires that the product $K_1.K_2.a_1.a_2$ equals one. When a_1, a_2 and/or K_1 are reduced below one, the gain factor K_2 must be larger than one. This will increase the quantisation noise due to $E_2(z)$, according to expression (4.49). E.g. when a stable input range of $0.7.V_{ref}$ is required for $x(k)$, the signal amplitude at the second integrator output can be $4.V_{ref}$, according to Fig. 3.27.c. In order to prevent an overload of the second stage, this signal has to scaled with a factor of 1/4, either by reducing a_1, a_2 or K_1 below unity. A gain factor K_2 equal to four is then required to obtain a cancellation of the noise due to $e_1(k)$. Due to this factor, the noise from $e_2(k)$ is increased with a factor of four and hence, the SNR is reduced with 12 dB. This corresponds to a resolution reduction of two bit.

A solution to overcome this problem is depicted in Fig. 4.37. [42]. Here, the second stage is a multi-bit Sigma-Delta converter, allowing a larger signal amplitude at its input. The effects of the DAC nonlinearities are significantly reduced by the Second-order noise shaping in the digital postprocessing circuit. The noise generated by these DAC nonidealities is negligible when the DAC Integral Nonlinearity satisfies the following expression:

$$INL < 1/4 \sqrt{\pi/OR} \tag{4.52}$$

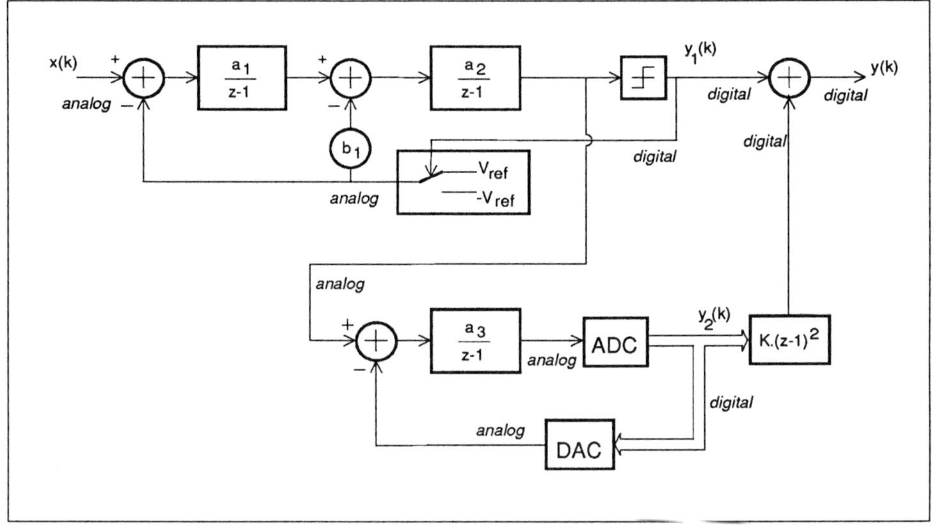

Fig. 4.37: a multi-bit MASH converter

4.6. SUMMARY

In this chapter, the problems involved with the design of higher-order Sigma-Delta modulators are discussed.

It is shown that a Sigma-Delta modulator with order larger than two is only conditionally stable: for some input signals and some initial conditions of the state variables, the loop is stable while for others, the state variables grow without bound. With computer simulations, stability conditions are derived for a Fourth-order modulator with clippers after the two last integrators. With a well-considered scaling of the signals at the integrator outputs, an unconditionally stable modulator is obtained.

Some alternatives for this higher-order converter topology are described in section 4.5.

Compared with standard Switched-Capacitor circuits, a Sigma-Delta modulator is more tolerant to circuit imperfections such as component tolerances, incomplete charge transfers, offset voltages, integrator leakages, clock feedthrough or thermal noise. The nonlinearity and the 1/f noise of the first integrator in Fig. 4.1. are the major signal corrupting circuit effects. Because of this tolerance for circuit imperfections, much larger clock frequencies can be achieved than in switched capacitor filters.

4.7. REFERENCES

[1]   S.A. TRETTER: "Introduction to Discrete-Time Signal Processing" - Wiley and Sons, New York, 1976

[2]   A. GELBE, W. VANDER VELDE: "Multiple-Input Describing Functions and Nonlinear System Design" - McGraw-Hill, New York, 1968

[3]   J. C. CANDY: "A Use of Double Integration in Sigma-Delta modulation" - *IEEE Trans. on Commun.* vol. COM-33, March 1985, pp. 249-258

[4]   W.L. LEE, C.G. SODINI - "A Topology for Higher Order Interpolative Coders", *Proc. of the ISCAS*, 1987, pp. 495-462

[5]   W. SANSEN, HUANG QIUTING, K. HALONEN: "Transient Analyisis of Charge transfer in SC Filters - Gain Error and Distortion", IEEE *J. Solid-State Circuits* vol. SC-22, no. 2, pp. 268-278, April 1987

[6]   P.E. FLEISCHER, A. GANESAN, K. LAKER: "Parasitic Compensated Switched Capacitor Filters" - *IEEE Trans. Circuits and Syst*, vol. CAS-27, pp. 237-244, April 1981

[7]   K. MARTIN, A.S. SEDRA: "Effects of the Op Amp Finite Gain and Bandwidth on the Performance of Switched-Capacitor Filters" - *IEEE Trans. Circuits Syst.*, vol. CAS-28, p. 822-829, Aug. 1981

[8]   B. DELSIGNORE *et al:* "A Monolithic 20b Delta-Sigma A/D Converter" - *ISSCC Digest of Techn. papers*, 1990, pp. 170-171

[9]   M. REBESCHINI *et al:* "A high-resolution CMOS Sigma-Delta A/D Converter with 320 kHz output rate" - *Proc. of the IEEE CICC*, 1989, pp. 6.1.1.-4.

[10]   R. KOCH *et al:* "A 12 bit Sigma-Delta analog-to-digital Converter with 15 MHz Clock rate" - *IEEE J. Solid-State Circuits* vol. SC-21 no.6, Dec. 1986, pp. 1003-1010

[11]   V. COMINO *et al:* "A First-Order Current-Steering Sigma-Delta Modulator" - *IEEE J. Solid-State Circuits* vol. SC-26 no.3, March 1991, pp. 167-183

[12]   F. OP 'T EYNDE, G. M. YIN, W. SANSEN: "A CMOS Fourth-Order 14b 500k-Sample/s Sigma-Delta ADC Converter" - *IEEE Int. Solid-State Circ. Conf, 1991 Dig. of Techn. Papers*, pp. 62-63

[13]   B. J. MCCAROLL, C. G. SODINI, H-S. LEE: "A High-Speed CMOS Comparator for use in an ADC" - *IEEE J. Solid-State Circuits*, vol. 23, pp. 159-165, Feb. 1988

[14]   P. NAUS *et al:* "Low Signal-level Distortion in Sigma-Delta modulators" - 84th Audio Eng. Soc. Convention, March 1988, Paris

[15]   J. KUO, R. DUTTON, B. WOOLEY: "MOS Pass transistor Turn-off Transient Analysis" - *IEEE Trans. Electr. Dev.* vol. ED-33 no. 10, Oct. 1986, pp. 1545-1555

[16]   B. J. SHEU, C. HU, "Modelling the Switched-induced Error Voltage on a Switched Capacitor" - *IEEE Trans. Circuits and Syst.* vol. CAS-30 no. 12, Dec. 1983, pp. 911-913

[17]   P. VANPETEGHEM, W. SANSEN: "Single versus complementary switches: a discussion of clock feedthrough in SC circuits" - Proc. *European Solid-State Circuits Conference* 1986, Delft, the Netherlands

[18]   P. VANPETEGHEM: "Accuracy and Resolution of switched capacitor circuits in MOS Technology" - Ph. D. Thesis, Katholieke Universitieit Leuven, 1986

[19]   K. HALONEN: "Low Power High Performance Switched Capacitor circuits for data-acquisition systems" - Ph. D. Thesis, Katholieke Universiteit Leuven, 1987

[20]   C.G. SHIH, P.R. GRAY: "Reference Refreshing Cyclic A-to-D and D-to-A Converters" - *IEEE J. Solid-State Circuits* vol. SC-21. no. 3, Aug. 1986 pp. 544-554

[21]   B. FURRER, W. GUGGENBUHL: "Noise Analysis of Sampled-Data Circuits" - *Proc. of the ISCAS*, 1981 pp. 860-863

[22]   C. GOBET, A. KNOB: "Noise analysis of Switched Capacitor networks" - *Proc. ISCAS*, 1981 pp.37-43

[23]   J. FISCHER: "Noise Sources and Calculation Techniques for Switched Capacitor Filters" - *IEEE J. Solid-State Circuits*, vol. SC-17, pp. 742-752, Aug. 1982

[24]   R.J. KANSKY: "Response of a Correlated Double Sampling Circuit to 1/f noise" - *IEEE J. Solid-State Circuits* vol. SC-15 no. 3, pp. 373-375, June 1980

[25]   C. HSIEH, P.R. GRAY *et al:* "Low-Noise Chopper-Stabilized Switched-Capacitor Filtering Technique" - *IEEE J. Sold-State Circuits*, vol. SC-16, Dec. 1981, pp. 708-715

[26]   E. VITTOZ: "The design of high-performance analog circuits on digital CMOS chips" - *IEEE J. Solid-State Circuits*, vol. SC-20, no. 3, June 1985, pp. 657-665

[27]   K. HALONEN, W. SANSEN: "Effect of current spikes in power supply rails on PSRR performance of SC-filter" - *Proc. ISCAS 1987*, pp. 64-67

[28]   P. VAN PETEGHEM: "On the Relationship between the PSRR and the Clock Feedthrough in SC Filters" - *IEEE J. Solid-State Circuits* vol. SC-23 no.4, Aug. 1988, pp. 997-1004

[29]   private communication with ir. D. LOGIE

[30]   M. BANU, J.M. KHOURY, Y. TSIVIDIS: "Fully Differential Operational Amplifiers with Accurate Output Balancing" - *IEEE J. Solid-State Circuits*, vol. SC-23, no. 6, Dec. 1988, pp.1410-1414

[31]   P. FERGUSON *et al:* "An 18b, 20KHZ Dual $\Sigma\Delta$ A/D Converter" - *IEEE 1991 ISSCC Digest of Techn. Papers*, pp. 68-69

[32]   J.C. CANDY: "A Use of double integration in Sigma-Delta modulation" - *IEEE Trans. Commun.*, pp. 294-258, March 1985

[33]   T. CATALTEPE, G. C. TEMES *et al*: "Digitally corrected multi-bit Sigma-Delta converters" - *IEEE Proc. Int. Symposium on Circuits and Systems*, pp. 647-650, May 1990

[34]   R. CARLEY: "A noise-shaping coder topology for 15+ bit converters" - *IEEE J. Solid-state Circuits*, pp. 267-273, April 1989

[35]   B. H. LEUNG: "Multibit Sigma-Delta A/D Converter Incorporating a Novel class of Dynamic Element Matching Techniques" - *IEEE Trans. on Circuits and Systems*, vol. CAS-39, no. 1, Jan. 1992, pp. 35-51

[36]   T. HAYASHI *et al*: "A multistage delta-sigma modulator, without double integrator loop" - *IEEE Int. Solid-state Circ. Conference, Digest of techn. Papers*, Febr. 1986, pp.182-183

[37]   Y. MASTUYA *et al*: "A 16-bit A/D conversion technology using triple integration noise shaping" - *IEEE J. Solid-state Circuits*, vol. SC-22, no. 6, Dec. 1987, pp. 921-929

[38]   M. REBESHINI *et al*: "A 16-bit 160 kHz CMOS A/D converter using Sigma-Delta modulation" - *proc. IEEE Custom Int. Circuits Conf*, May 1989, pp. 6.1.1-4.

[39]   L. LONGO, M. A. COPELAND: "A 13-bit ISDN-band ADC using two-stage third-order noise shaping" - *Proc. Custom Int. Circuits Conf*, June 1988, pp. 21.1-4

[40]   M. REBESHINI *et al*: "A High-resolution CMOS Sigma-Delta A-to-D converter with 320 kHz output rate" *Proc. IEEE Symp. Circuits and Systems*, May 1989, pp. 246-249

[41]   D. B. RIBNER *et al*: "16b Third-order Sigma-Delta modulator with reduced Sensitivity to Nonidealities" - *IEEE Int. Solid-state Circ. Conference, Digest of techn. Papers*, Febr. 1991, pp. 66-67

[42]   B. P. BRANDT, B. A. WOOLEY: "A CMOS Oversampling A/D Converter with 12b resolution at conversion rates above 1MHz" - *IEEE Int. Solid-state Circ. Conference, Digest of techn. Papers*, Febr. 1991, pp. 64-65

APPENDIX 4.A. SOME DISTORTION GENERATION MECHANISMS IN A SIGMA-DELTA ADC

In Fig. 4.1, the principle schematic of a Sigma-Delta ADC is shown. $x(k)$ is the analog ADC input voltage, $y(k)$ is the one-bit PDM signal (equal to one or minus one) and $d(k)$ is the multi-bit digital output. Assume that the first integrator can be described with the following nonlinear difference equation:

$$v_1(k+1) = v_1(k) + a_{11} \cdot z_1(k) + a_{12} \cdot z_1(k)^2 \qquad (4.4)$$

where $z_1(k)$ is the integrator input voltage, equal to

$$z_1(k) = x(k) - y(k) \cdot V_{ref} \qquad (4.29)$$

The objective of this appendix is to calculate the nonlinear DC transfer characteristic from the input $x(k)$ to the output $d(k)$. From this nonlinear characteristic, the harmonic distortion due to the nonlinear integrator gain can be retrieved according to the techniques discussed in chapter 2.

In order to determine the nonlinear DC transfer characteristic, assume that a DC input signal X is applied to the ADC input. The PDM signal $y(k)$ can be either one or minus one. Let us denote the probability that $y(k)$ equals one with the symbol "p":

$$p = \frac{\text{the number of bit equal to one in a long output sequence}}{\text{the number of ones + the number of minus ones}} \qquad (4.53)$$

E.g. when all the bits in the PDM signal are one, p equals 1. When all the bits in the PDM signal are minus one, p is 0. The probability that $y(k)$ equals minus one is $1-p$.

Since for a stable modulator, the output of the first integrator cannot grow without limit, the average increment of $v_1(k)$ has to equal zero. This implies with expressions (4.4) and (4.52) that

$$p \cdot [a_{11} \cdot (X-Vref) + a_{12} \cdot (X-Vref)^2] +$$
$$(1-p) \cdot [a_{11} \cdot (X+V_{ref}) + a_{12} \cdot (X+V_{ref})^2] = 0 \qquad (4.54)$$

$$\text{or,} \quad p = \frac{a_{11} \cdot (X+V_{ref}) + a_{12} \cdot (X+V_{ref})^2}{2 \cdot a_{11} \cdot V_{ref} + 4 \cdot a_{12} \cdot X \cdot V_{ref}} \qquad (4.55)$$

The DC value of the digital output $d(k)$ is equal to the DC value of the PDM signal, which is given by

$$DC(y(k)) = p \cdot 1 + (1-p) \cdot (-1) \qquad (4.56)$$

From (4.54) and (4.56), the relation between X and the DC value of d(k) can be obtained:

$$DC(d(k)) = \frac{a_{12}}{a_{11}}.V_{ref} + \frac{X}{V_{ref}} - \frac{a_{12}}{a_{11}}.\frac{X^2}{V_{ref}} + small\ terms \qquad (4.57)$$

The first term of this expression represents an output refered offset, the second term is the linear ADC gain and the third term represents the second order nonlinear gain.

When a sinusoidal voltage with amplitude A_{in} is applied to the ADC input, the second harmonic distortion can be calculated from expressions (4.57) and (2.9):

$$HD_2 = \frac{1}{2}.\frac{a_{12}}{a_{11}}.A_{in} \qquad (4.5)$$

This result is a bit surprising: although the first integrator is within a feedback loop, its nonlinearity is not suppressed by the loop gain. This is due to the unusual nature of the feedback signal y(k): the signal is represented by a pulse density instead of a voltage.

Another circuit imperfection that causes distortion is due to the finite settling times of the reference voltage buffers in Fig. 4.14.a. The charge transferred to C_2 (see Fig. 4.14.a.) during ϕ_2 equals

$$Q = (1-\varepsilon^+)\cdot C_1.(V_{ref}-V_{in}) \qquad if\ y(k)=1 \qquad (4.17.a)$$

$$= (1-\varepsilon^-).C_1.(-V_{ref}-V_{in}) \qquad if\ y(k)= -1 \qquad (4.17.b)$$

In this expression, ε^+ and ε^- are the charge transfer errors during ϕ_2, respectively when S_1 is closed and when S_2 is closed. Due to differences between the buffer amplifiers that buffer the two reference voltages V_{ref} and -Vref, ε^+ might be different from ε^-. It is shown here that this difference causes distortion.

Since the integrator output cannot grow without limit, the average charge has to equal zero:

$$p.(1-\varepsilon^+).C_1.(V_{ref}-V_{in}) + (1-p).(1-\varepsilon^-).C_1.(-V_{ref}-V_{in}) = 0 \qquad (4.58)$$

where p is the probability that y(k) equals one. From (4.57), (4.58) and (2.9), the harmonic distortion at the ADC output can be calculated:

$$HD_2 = \frac{1}{4}.\frac{X}{V_{ref}}.(\varepsilon^--\varepsilon^+) \qquad (4.18)$$

A difference between ε^+ and ε^- causes harmonic distortion, even when the charge transfer errors are signal independent. Again, this result is surprising: in standard sampled-data circuits, signal independent charge transfer errors cause no harmonic distortion [5].

5 THE PRACTICAL IMPLEMENTATION OF SIGMA-DELTA D-TO-A CONVERTERS

5.1. INTRODUCTION

A signal conversion from an analog to a digital representation requires a sampling and a quantisation (see Fig. 3.1.a.). The former limits the signal bandwidth while the latter causes quantisation noise. Therefore, an ADC degrades the signal quality. In a DAC on the other hand, the digital input signal is converted to an analog quantity and subsequently transformed into a continuous-time, discrete-amplitude signal by a hold operation (see Fig. 3.1.b.). Provided that the conversion to analog is perfectly linear and that the sampling times are exactly equidistant, the DAC causes no additional signal degradation. The noise at the analog output is entirely due to the discrete amplitude nature of the digital input signal.

However, practical D-to-A converters suffer from circuit nonidealities such as device mismatches, component noise, device nonlinearities or clock jitter. Because of these, the DAC causes an additional signal degradation. This degradation has to be below a half bin, otherwise the input signal LSB is not significant. Hence, the relation between the INL and the SNR of a DAC is the same as for an ADC (see expression (3.8)).

Fig. 5.1. shows the basic schematic of a Sigma-Delta DAC. With a digital interpolator, the sampling rate is increased from f_d to f_s. The Oversampling Ratio of the resulting signal is larger than one. The word length is then reduced to one bit with a digital Sigma-Delta modulator. A large amount of quantisation noise is added to the signal in this step but most of the noise power is outside the signal band. The one bit PDM signal is converted to an analog quantity, for instance by switching between two reference voltages. The high-frequency noise is removed with an analog lowpass filter, yielding the required analog output signal. The motivation of this rather complex principle is the same as for an ADC: the conversion linearity is independent of component mismatches.

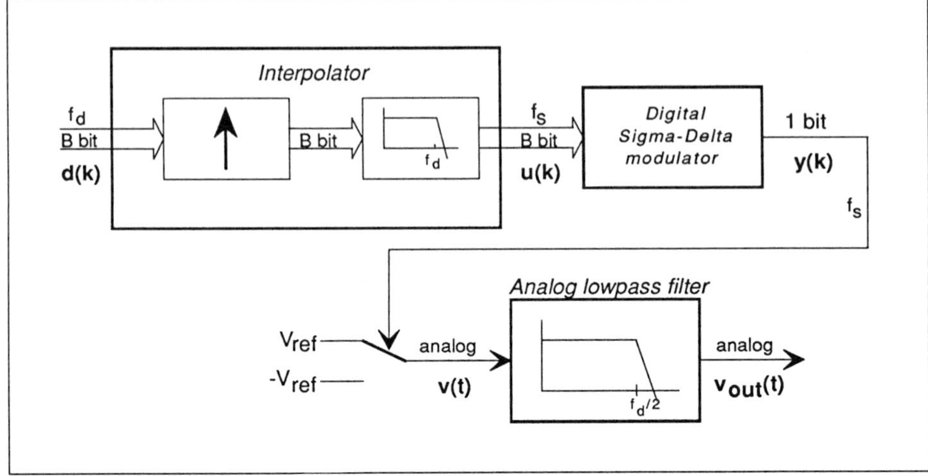

Fig. 5.1: The basic principle of a Sigma-Delta DAC

Although the analog part of this schematic looks very simple - just two reference voltages, two switches and a (passive) lowpass filter - the practical implementation is not as obvious as it seems at first sight. Besides signal degradation due to the in-band quantisation noise introduced by the Sigma-Delta modulator, the DAC also suffers from component noise, noise due to clock jitter and from circuit nonlinearities. Just because of these circuit imperfections, the analog reconstruction is the most difficult analog building block in a DSP system with Sigma-Delta interfaces.

The clock frequency f_s is much larger than the input signal frequency. Hence, when simulating the circuit behaviour with a circuit-level simulator such as SPICE, a large number of switching transients have to be simulated to obtain one signal period. This results in an unrealistic calculation effort and in an enormous number of simulated time points. It is not easy to obtain the in-band signal power spectrum from this large amount of data. Therefore, circuit-level simulations are not feasible. The implications of analog circuit nonidealities can only be obtained from analytical calculations or from measurements.

The influence of **component noise** in the analog part of the circuit can be analysed with standard noise calculation techniques for sampled-data systems. Although this noise can be very severe, it is a well-known phenomenon [1] and therefore not repeated in this text.

The main advantage of the Sigma-Delta principle is the independence of component mismatches. In section 5.2, it is shown that this advantage is obtained at the cost of a higher sensitivity to **clock jitter**. Expressions for the jitter noise are presented in this section.

In most analog circuits, the **harmonic distortion** increases with the signal amplitude. In a Sigma-Delta DAC however, the PDM signal y(k) varies between one and minus one and its amplitude is signal independent. The input signal d(k) is represented by the pulse density of y(k). Because of this unconventional signal representation, it is not obvious which circuit nonlinearities cause harmonic distortion. This problem has been brought up only in a few papers up to now, for some specific circuits [2] [3]. In this chapter, a general technique is demonstrated to calculate the harmonic distortion due to circuit imperfections.

In section 5.2, it is shown that the circuit of Fig. 5.1. is inherently nonlinear. Especially for those high-performance applications that require a Sigma-Delta modulator, this circuit cannot be used because it generates too much harmonic distortion.

In section 5.3, an alternative circuit is discussed. Instead of switching between two reference voltages, the analog output is formed by a current switched between two opposite values. This approach merely eliminates the nonlinearities of the previous circuit but the sensitivity to clock jitter is still too large. A design example is discussed, achieving 12 bit resolution for a signal bandwidth of 80 kHz.

In section 5.4, the principle of a Switched-Capacitor DAC is discussed. Here, the PDM signal is converted to a charge placed on a capacitor. This circuit is insensitive to clock jitter. However, a high speed, low-noise, very linear amplifier with a large slew rate is required to obtain reasonable distortion specifications.

5.2. A VOLTAGE DRIVEN DAC

In Fig. 5.2.a, the analog part of the Sigma-Delta DAC is repeated as it is shown in Fig. 5.1. It is illustrated here that this circuit is inherently nonlinear and very sensitive to clock jitter. Practical implementations either are limited to a resolution of 12 bit [4] or are built with discrete precision components [3].

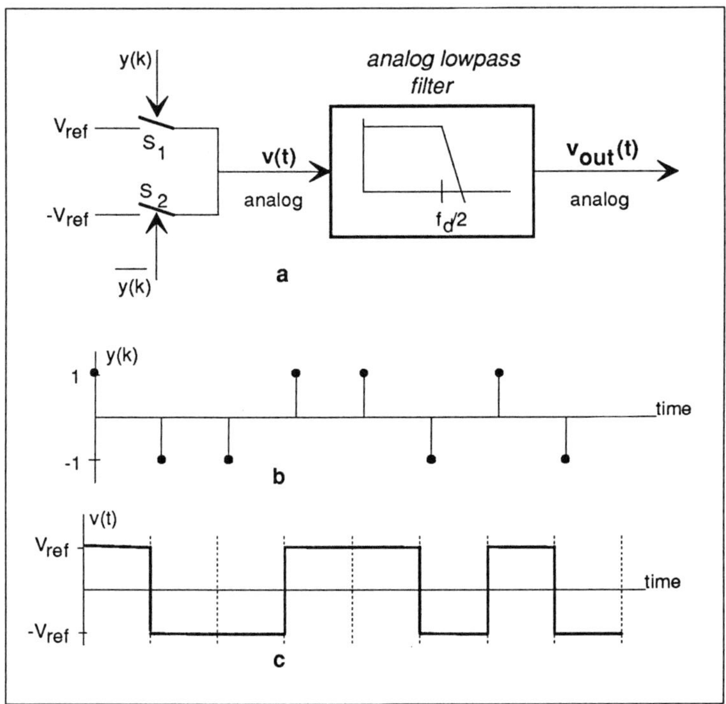

Fig. 5.2: The analog part of a voltage driven Sigma-Delta DAC

The ideal transfer function is analysed first. Afterwards, the distortion characteristics and the influence of clock jitter are discussed.

5.2.a. The Signal Transfer

When a digital signal d(k) is applied to the input in Fig. 5.1, the sampling frequency is increased first from the Nyquist rate f_d to a much higher rate f_s. For in-band frequencies, the spectrum of u(k) equals the spectrum of d(k):

$$U(j\omega) = D(j\omega) \tag{5.1}$$

In the Sigma-Delta modulator, the word length is truncated to one bit by adding quantisation noise E(jω) to the signal:

$$Y(j\omega) = U(j\omega) + E(j\omega) \tag{5.2}$$

This modulator is similar to the circuits discussed in the previous chapters but now the integrators and summators are formed with digital accumulators (see Fig. 3.23.b.). Since the attention is devoted in this chapter to the analog circuitry of the DAC, no assumptions are made concerning the exact shape of the quantisation noise power spectrum or on the Sigma-Delta modulator order. The only restrictions are that the in-band quantisation noise power is small and that its DC value is zero.

In the analog reconstruction, the PDM signal is converted to an analog voltage. When the switching transients of v(t) are assumed infinitely fast, the waveforms in Fig. 5.2.c. are rectangular. The DAC performs an ideal hold operation. The analog signal v(t) is constant during the k-th clock cycle and given by:

$$v(t) = V_{ref} \qquad \text{if } y(k) = 1 \tag{5.3.a}$$

$$v(t) = -V_{ref} \qquad \text{if } y(k) = -1 \tag{5.3.b}$$

This can be reformulated in a closed expression:

$$v(t) = y(k).V_{ref} \tag{5.3.c}$$

or, in the frequency domain,

$$V(j\omega) = Y(j\omega).V_{ref}.T.\frac{\sin(\omega T/2)}{\omega T/2}.\exp(-j\omega T/2)$$

$$= [D(j\omega) + E(j\omega)].V_{ref}.T.\frac{\sin(\omega T/2)}{\omega T/2}.\exp(-j\omega T/2) \tag{5.4}$$

where T is the clock period. The frequency dependence is due to the hold operation (see expression (3.9)). As demonstrated in section 3.5, the signal attenuation due to the hold operation is small for in-band signals. Hence, expression (5.4) can be approximated with

$$V(j\omega) = [D(j\omega) + E(j\omega)].V_{ref}.T \tag{5.5}$$

With a lowpass filter, the out-band quantisation noise is removed. $V_{out}(t)$ contains the requested analog equivalent of the digital input plus a small in-band quantisation noise signal.

In contrast with a classical multi-level DAC based on binary weighted arrays, no matching analog components are required. Mismatches between the two reference voltages are not much of a problem. Assume for instance that the reference voltages differ from their nominal value as shown in Fig. 5.3. The analog signal v(t) can now be expressed as

$$v(t) = V_{ref} + \Delta V_{ref1} \qquad \text{if } y(k) = 1 \text{ and} \qquad (5.6.a)$$

$$v(t) = -V_{ref} + \Delta V_{ref2} \qquad \text{if } y(k) = -1 \qquad (5.6.b)$$

This can be reformulated in a closed form. Note that

$$\frac{1+y(k)}{2} \quad \text{equals one when y(k) is one and equals zero when y(k) is minus one} \qquad (5.7.a)$$

$$\frac{1-y(k)}{2} \quad \text{equals zero when y(k) is one and equals one when y(k) is minus one} \qquad (5.7.b)$$

Hence, expressions (5.6) are equivalent with

$$v(t) = (V_{ref}+\Delta V_{ref1}) \cdot \frac{1+y(k)}{2} + (-V_{ref}+\Delta V_{ref2}) \cdot \frac{1-y(k)}{2} \qquad (5.8.a)$$

$$= (V_{ref}+ \frac{\Delta V_{ref1}-\Delta V_{ref2}}{2}) \cdot y(k) + \frac{\Delta V_{ref1}+\Delta V_{ref2}}{2} \qquad (5.8.b)$$

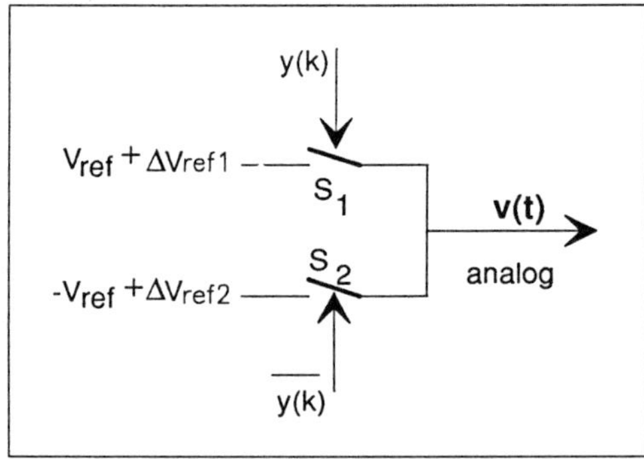

Fig. 5.3: A DAC with tolerances on the reference voltages

Compare this with formula (5.3.c). As can be seen, the deviations ΔV_{ref1} and ΔV_{ref2} cause a gain error and an offset, but no nonlinearity. These effects can be accepted in most applications.

Clock feedthrough is not a major problem either: at the moment that switch S_1 in Fig. 5.2.a. opens, the voltages at both switch terminals equal V_{ref}. The charge injected by this switch is always the same. In a similar way, the charge injected by S_2 is invariant. As pointed out in chapter 4, these charges cause an offset or a gain error but no harmonic distortion.

5.2.b. The distortion characteristics

The analog signal v(t) in Fig. 5.1. is switched between two reference voltages. Therefore, v(t) is a large signal even when the digital input signal amplitude is small. When the switching transients are infinitely fast and when the lowpass filter is perfectly linear, only the low-frequency content of this signal appears at the output.

In practice however, since v(t) is a large signal, even small circuit nonlinearities cause a considerable harmonic distortion. Moreover, v(t) contains a large amount of high-frequency quantisation noise. Circuit nonlinearities cause an intermodulation of this quantisation noise. Intermodulation of two large high-frequency noise components at frequencies close to each other yields an important in-band noise component that can pass the lowpass filter and degrade the Signal-to-Noise Ratio.

In Fig. 5.4.a, an example of the waveform v(t) after the analog reconstruction is shown. Since only the low-frequency content of this signal passes the lowpass filter, this signal can be approximated by its average value V(k) during each clock cycle (see Fig. 5.4.b.). The low-frequency contents of the two signals in Fig. 5.4. are approximately the same.

Due to capacitive loading, it takes some time for v(t) to switch from one reference voltage to the other. When y(k) and y(k-1) are both one, the average voltage V(k) during the k-th clock period equals V_{ref} (see the fifth clock period in Fig. 5.4.). Similarly, V(k) equals minus V_{ref} when y(k) and y(k-1) are both minus one (e.g. the third period in Fig. 5.4.). But when y(k) and y(k-1) are of opposite sign, the analog voltage has to switch from the one reference voltage to the other. The average, V(k), is not exactly equal to one of the reference voltages. This implies that V(k) during the k-th clock period is not only function of the current PDM value, y(k), but also of the previous one. It is shown here that this causes distortion.

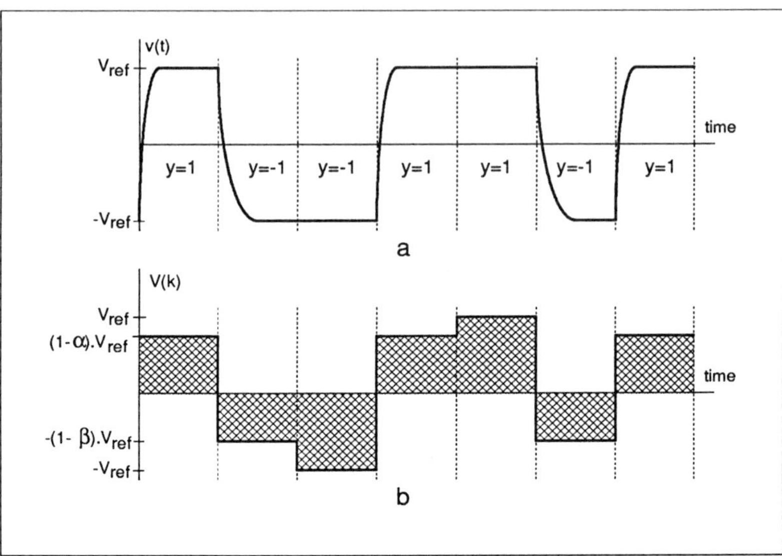

Fig 5.4: a) A typical waveform for v(t) and b) its average
V(k) over one clock cycle

Let us introduce the following notations (see Fig. 5.4.b.):

$$
\begin{aligned}
V(k) &= V_{ref} & \text{if } y(k-1) &= 1 \text{ and } y(k) = 1 \\
&= (1-\alpha) \cdot V_{ref} & \text{if } y(k-1) &= -1 \text{ and } y(k) = 1 \\
&= -V_{ref} & \text{if } y(k-1) &= -1 \text{ and } y(k) = -1 \\
&= -(1-\beta) \cdot V_{ref} & \text{if } y(k-1) &= 1 \text{ and } y(k) = -1
\end{aligned}
\tag{5.9}
$$

α and β are small positive numbers. An example will be given below. With the use of expressions (5.7), the expressions (5.9) can be reformulated in a closed form:

$$
V(k) = \frac{1+y(k-1)}{2} \cdot \frac{1+y(k)}{2} \cdot V_{ref} + \frac{1-y(k-1)}{2} \cdot \frac{1+y(k)}{2} \cdot (1-\alpha) \cdot V_{ref}
$$

$$
- \frac{1+y(k-1)}{2} \cdot \frac{1-y(k)}{2} \cdot (1-\beta) \cdot V_{ref} - \frac{1-y(k-1)}{2} \cdot \frac{1-y(k)}{2} \cdot V_{ref}
\tag{5.10}
$$

$$
= \{ (\beta-\alpha) + (\alpha+\beta) \cdot y(k-1) + (4-\alpha-\beta) \cdot y(k) +
$$

$$
(\alpha-\beta) \cdot y(k) \cdot y(k-1) \} \cdot V_{ref}/4
\tag{5.11}
$$

Due to the finite speed of the switching transients, the circuit "remembers" the previous PDM value. When the switching transients are different for a positive and for a negative transition, α and β are different and expression (5.11) contains a term proportional with the product of y(k) and y(k-1). The low-frequency content of this signal will pass the lowpass filter and appear at the output.

When a quasi-static sinusoidal digital signal is applied at the DAC input, y(k) and y(k-1) both consist of a sinusoidal signal plus quantisation noise. Due to the product in the last term of (5.11), a second harmonic component of the sinusoidal signal is generated. The second harmonic distortion can be calculated from expressions (5.11) and (2.9):

$$HD_2 \;=\; \frac{1}{8} . (\alpha-\beta) . A \tag{5.12}$$

where A is the digital input signal amplitude (between zero and one).

The product of y(k) and y(k-1) causes also an intermodulation of the high-frequency quantisation noise, yielding low-frequency intermodulation products. The resulting SNR degradation depends upon the exact shape of the high-frequency quantisation noise, thus upon the modulator order. A safe upper limit for the in-band noise power is given by:

$$N \;=\; (\alpha-\beta)^2 . V_{ref}^2 / 32 \quad \text{and} \quad SNR \;=\; -10 . \log[((\alpha-\beta)^2 / 16] \;\; dB \tag{5.13}$$

E.g. when $(\alpha-\beta)$ equals 0.01, the second harmonic distortion according to (5.12) is -58 dB for a full scale input. The SNR according to (5.13) is only 52 dB. This illustrates that the circuit of Fig. 5.2, in its original form, is not useful for applications with a high dynamic range.

The signal degradation expressed by formulas (5.12) and (5.13) is due to the fact that the circuit "memorises" the previous PDM value. A simple solution for this problem is demonstrated in Fig. 5.5. [5]. With a third switch, v(t) is connected to the ground voltage between two pulses. As can be seen from this drawing, the waveforms during each clock period are only function of the current PDM value and independent of y(k-1). All the positive pulses are now identical, and so are the negative ones. According to the symbols in Fig. 5.5.c, the average voltage V(k) can be denoted as:

$$
\begin{aligned}
V(k) \;&=\; \alpha . V_{ref} && \text{if } y(k) \;=\; 1 \\
&=\; -\beta . V_{ref} && \text{if } y(k) \;=\; -1
\end{aligned}
\tag{5.14}
$$

α and β are in the order of 0.5. With the use of expressions (5.7), this can be reformulated as:

$$V(k) = \alpha \cdot V_{ref} \cdot \frac{1+y(k)}{2} - \beta \cdot V_{ref} \cdot \frac{1-y(k)}{2} \qquad (5.15.a)$$

$$= \{ \frac{\alpha-\beta}{2} + \frac{\alpha+\beta}{2} \cdot y(k) \} \cdot V_{ref} \qquad (5.15.b)$$

This expression contains no second-order nonlinear term. Therefore, the circuit of Fig. 5.5. causes no harmonic distortion or quantisation noise intermodulation.

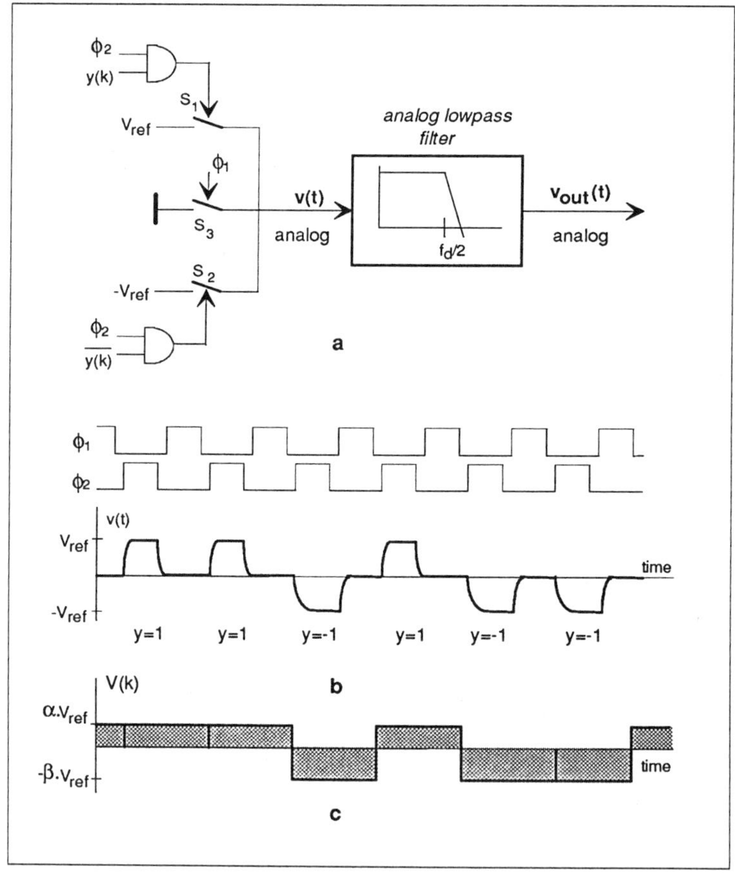

Fig. 5.5: An improved voltage driven DAC without "memory"

This conclusion can be generalised. y(k) is a PDM signal. Its pulse density is linearly proportional with the input signal d(k): the higher the DAC input signal value, the more ones (and the less minus ones) there are in the PDM bit stream. In the same way, the output signal (v_{out}(t) in Fig. 5.5.a.) is constructed from two kinds of pulses (positive ones and negative ones, see Fig. 5.5.b.). The output signal value is determined by the relative occurrence of these pulses: the more positive pulses in v(t), the higher the output signal. **When all the pulses of the same sign in Figs. 5.4.a. and 5.5.b. are identical**, the average voltage V(k) is of the same form as expressions (5.15) and contains no nonlinearity.

The exact shape of these pulses is not important. For instance, when the switch on-resitances in Fig. 5.5.a. are nonlinear, the switch transients in Fig. 5.5.b. are not exactly of an exponential form. In classical sampled-data circuits (e.g. Switched-Capacitor filters) this would result in harmonic distortion. In a Sigma-Delta DAC, this nonlinearity is not important because all the pulses with the same sign have identical non-exponential shapes.

The nonlinearity of the lowpass filter input stage in Fig. 5.5.a. is a second possible mechanism that generates distortion or causes quantisation noise intermodulation. Since v(t) is switched between V_{ref} and $-V_{ref}$, the signal amplitude at the filter input is always large, even when the DAC input amplitude is small. When the lowpass filter is built with active components, this large signal causes linearity problems. Hence, it is good practice [2] [3] [4] to include a passive lowpass filter stage immediately after the analog switches. This is illustrated in Fig. 5.6.a. In this way, most of the high-frequency quantisation noise is removed and the lowpass filter input signal amplitude is reduced. The RC cut-off frequency is somewhat larger than the signal bandwidth.

However, this RC filter causes additional linearity problems. Consider a positive pulse, as shown in Fig. 5.6.b. (the situation is similar for negative pulses). After the switching transient, v(t) in Fig. 5.6.a. reaches a stable voltage, equal to

$$\frac{V_{ref}.R + w(t).R_{S1}}{R + R_{S1}}, \text{ where } R_{S1} \text{ is the on-resistance of } S_1.$$

As can be seen, the waveform of this pulse is a function of the voltage over the capacitor, w(t). This voltage is determined by the charge stored on the filter capacitor during previous clock periods, thus by previous values of the PDM signal. Hence, not all the pulses of the same sign are identical and distortion will occur.

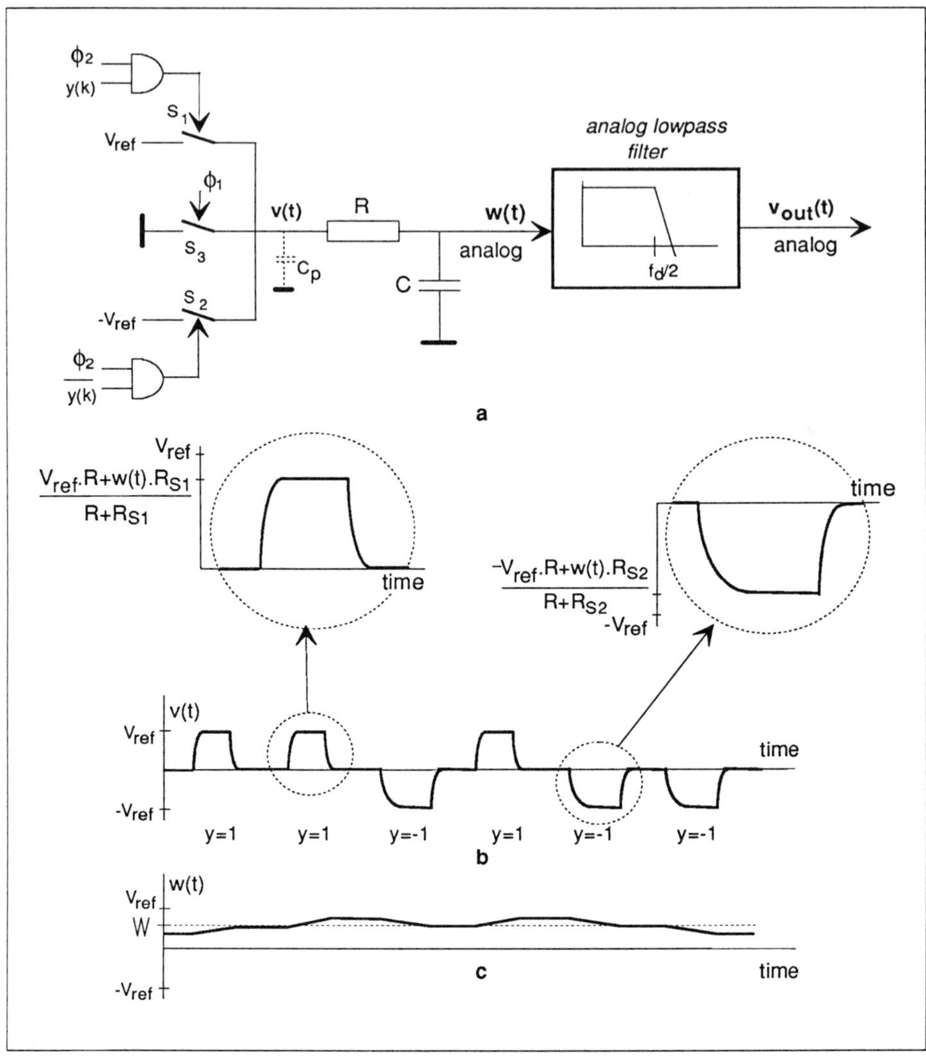

*Fig. 5.6: A voltage driven DAC with a passive lowpass
filter stage*

This harmonic distortion can be estimated by calculating the nonlinear DC transfer characteristic from the DAC input d(k) to the output $v_{out}(t)$. Consider the circuits of Fig. 5.1. and Fig. 5.6. and assume that a constant digital word D (between minus one and one) is applied to the input in Fig. 5.1. In the Sigma-Delta modulator, the digital input is converted to the PDM signal y(k). Let us denote the pulse density of y(k) with the symbol "p". This pulse density is function of the DAC input signal D.

As pointed out in section 3.4.d, the quantisation noise of a Sigma-Delta modulator contains no DC component. Hence, the DC component of y(k) equals the DC component of the DAC input:

$$p.1 + (1-p).(-1) = D \qquad (5.16)$$

$$\text{or,} \quad p = \frac{1+D}{2} \quad \text{and} \quad 1-p = \frac{1-D}{2} \qquad (5.17)$$

E.g. when the DAC input word D is one, y(k) is always one (p=1). When D is minus one, y(k) is always minus one and thus never equal to one (p=0). When D is zero, y(k) oscillates between one and minus one. Half of the samples of y(k) are one, the other half is minus one (p=0.5).

If we assume in first order that the parasitic capacitance C_p is zero, the switching transients in Fig. 5.6.b. are infinitely fast. The charge stored on the capacitor C during the k-th clock cycle is then equal to:

$$Q(k) = \frac{V_{ref}-W}{R+R_{S1}}.T_{on} \qquad \text{if} \quad y(k) = 1 \qquad (5.18.a)$$

$$= \frac{-V_{ref}-W}{R+R_{S2}}.T_{on} \qquad \text{if} \quad y(k) = -1 \qquad (5.18.b)$$

where W is the average of w(t) (see Fig. 5.6.c.).

Since the voltage over the capacitor C in Fig. 5.6.a. cannot grow without limit, the long-term average charge stored on this capacitor has to equal zero. Hence,

$$p.\frac{V_{ref}-W}{R+R_{S1}}.T_{on} + (1-p)\frac{-V_{ref}-W}{R+R_{S2}}.T_{on} = 0 \qquad (5.19)$$

R_{S1} and R_{S2} are the switch on-resistances of S_1 and S_2 respectively and T_{on} is the switch on-time. The output resistances of the reference voltage sources are in series with the switch on-resistances and can be included in R_{S1} and R_{S2}.

From expression (5.19), with the use of (5.17), the DC-transfer from D to W can be calculated:

$$W = \frac{(R_{S1}-R_{S2}) + D.(2.R+R_{S1}+R_{S2})}{(2.R+R_{S1}+R_{S2}) + D.(R_{S2}-R_{S1})}.V_{ref} \qquad (5.20)$$

$$= \{ \ \frac{R_{S1}-R_{S2}}{2.R+R_{S1}+R_{S2}} + D + \frac{R_{S1}-R_{S2}}{2.R+R_{S1}+R_{S2}}.D^2 + \ ... \ \}.V_{ref} \qquad (5.21)$$

The first term of this expression represents an offset, the second term is the linear gain and the last term is the second-order nonlinear gain. As can be seen, when the switch on-resistances R_{S1} and R_{S2} are different, the DC component of w(t) is a nonlinear function of the DAC input signal D. This DC component can pass the lowpass filter in Fig. 5.6.a. and appears at the DAC output. Hence, the DAC DC transfer function is nonlinear.

When a quasi-static sinusoidal digital signal with amplitude A (between zero and one) is applied to the DAC input, a second harmonic is generated. Expressions (5.21) and (2.9) yield:

$$HD_2 = \frac{1}{2}.\frac{R_{S1}-R_{S2}}{2.R+R_{S1}+R_{S2}}.A \approx \frac{1}{4}.\frac{R_{S1}-R_{S2}}{R}.A \qquad (5.22)$$

This result is remarkable: even when **all the analog components are linear**, a difference between the switch on-resistances R_{S1} and R_{S2} causes distortion. E.g. when the switch on-resistances are ten times smaller than the resistor R and with 5% mismatch between the switches, the second harmonic distortion is -58 dB for a full-scale input. This corresponds with a resolution of less than 10 bit.

In the same way [6], it can be shown that a difference between the output impedances of the reference voltage sources or a difference between the switch on-times cause harmonic distortion.

Up to now, it is assumed that C_p in Fig. 5.6.a. is zero. Also, the switch on-resistances are assumed linear. With a more involved calculation [6], it can be shown that a non-zero C_p or nonlinear switch on-resistances cause additional distortion.

Although the Sigma-Delta principle is theoretically insensitive for component mismatches, expression (5.22) illustrates that mismatches between the switch on-resistances cause harmonic distortion. This is the most severe distortion generation mechanism in a voltage driven DAC because there is no simple solution. Increasing the switch transistor widths to decrease the ratios R_{S1}/R and R_{S2}/R is no use: the parasitic nonlinear switch capacitance (C_p in Fig. 5.6.a.) increases as well. This causes additional distortion. A differential approach to suppress the second harmonic distortion will neither yield any improvement: since mismatches between switches are a random phenomenon, the mismatches are not equal in the two sides of a differential circuit and the second harmonic will not disappear.

In section 5.3, an alternative approach is discussed. It is pointed out that this eliminates the distortion due to mismatches.

5.2.c. The signal degradation due to clock jitter

Besides distortion, clock jitter is an other effect that degrades the signal quality. In Fig. 5.7.a, the waveform v(t) generated by the circuit of Fig. 5.6. is depicted. The average voltage during the k-th clock cycle equals approximately

$$V(k) = y(k) \cdot V_{ref} \cdot T_{on}/T \tag{5.23}$$

where T_{on} is the switch on-time and T is the clock period. Due to clock jitter, T_{on} can deviate form its nominal value, resulting in an error signal. This is illustrated in Fig. 5.7.b. The average voltage V(k) is changed with an amount:

$$\Delta V(k) = y(k) \cdot V_{ref} \cdot \Delta T_{on}/T \tag{5.24}$$

This error signal causes noise. If ΔT is assumed random with a white spectrum, the in-band noise power can be calculated as

$$N = 20 \cdot \log \left(\frac{\sigma(\Delta T)}{T} \cdot \frac{1}{\sqrt{2 \cdot OR}} \cdot V_{ref} \right) \text{ dB} \tag{5.25}$$

and

$$SNR = -20 \cdot \log \left(\frac{\sigma(\Delta T)}{T_{on}} \cdot \frac{1}{\sqrt{OR}} \right) \text{ dB} \tag{5.26}$$

In these expressions, $\sigma(\Delta T)$ is the standard deviation of the error ΔT. E.g. when the OR equals 100, with a clock frequency of 20 MHz, with T_{on} equal to T/2 and with a clock jitter $\sigma(\Delta T)$ of 10 psec, the SNR is 88 dB. This corresponds with a resolution of 14 bit.

Note that a similar problem exists in a classical multi-level DAC. In Fig. 5.7.c, a typical output of a classical DAC is depicted. Clock jitter causes an error signal as illustrated in Fig. 5.7.d. The error on V(k) can be calculated as:

$$\Delta V(k) = [y(k)-y(k-1)] \cdot V_{ref} \cdot \Delta T/T \tag{5.27}$$

Compare this expression with equation (5.24). For a slowly varying input signal, the difference between y(k) and y(k-1) is in the order of a few LSB, thus less than one (see Fig. 5.7.d.). Moreover, for the same signal bandwidth, the clock period T of a Sigma-Delta DAC is a factor OR smaller than that of a Sigma-Delta DAC. Hence, it can be concluded from (5.24) and (5.27) that the noise due to clock jitter is much larger for a Sigma-Delta DAC than for a multi-level DAC.

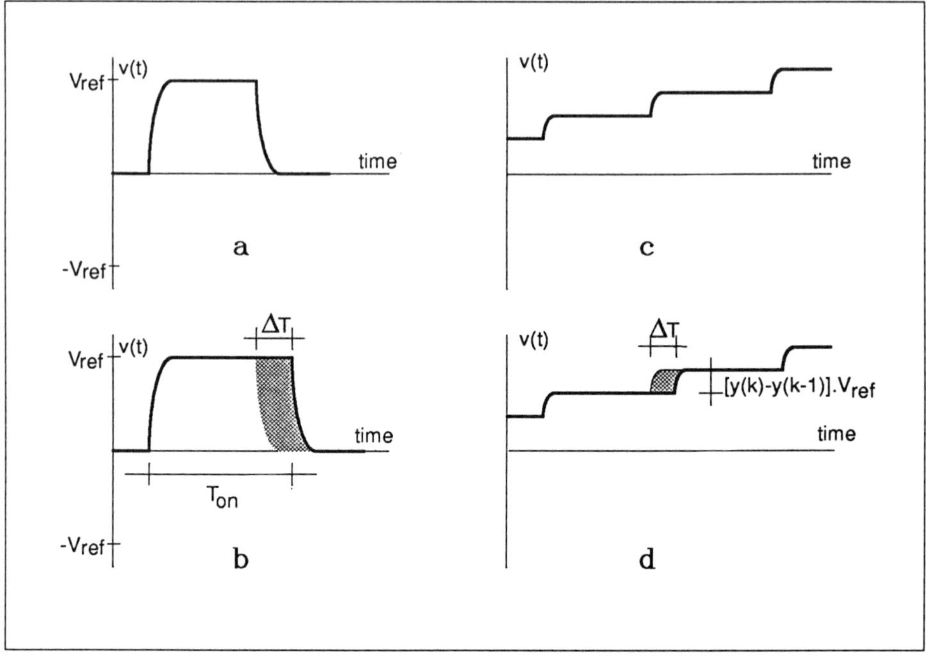

Fig. 5.7: Typical output waveforms of a DAC:
a) the signal v(t) of a Sigma-Delta DAC
b) the same voltage, but now disturbed by clock jitter
c) the output of a multi-level DAC
d) the same signal, but now disturbed by clock jitter

5.2.d. Conclusions

Due to analog circuit imperfections, practical implementations of the voltage driven DAC depicted in Fig. 5.1. suffer from linearity problems and from clock jitter. Therefore, the resolution is limited to about 10 bit. In the next section, an alternative approach is discussed that eliminates the nonlinearity problems.

5.3. A CURRENT DRIVEN DAC

5.3.a. The Signal Transfer

Instead of constructing the analog voltage v(t) by switching between two voltages, it is also possible to switch between two currents. This is illustrated in Fig. 5.8.a. During the clock phase ϕ_1, the switches are both open and the current i(t) is zero. During ϕ_2, the current equals I_{ref} or $-I_{ref}$, depending on the value of y(k). With an RC filter, most of the quantisation noise is filtered out in order to avoid a large signal with fast transients at the lowpass filter input. With an expression similar to (5.9), it can be shown that mismatches between the current source values cause only a gain factor or an offset voltage.

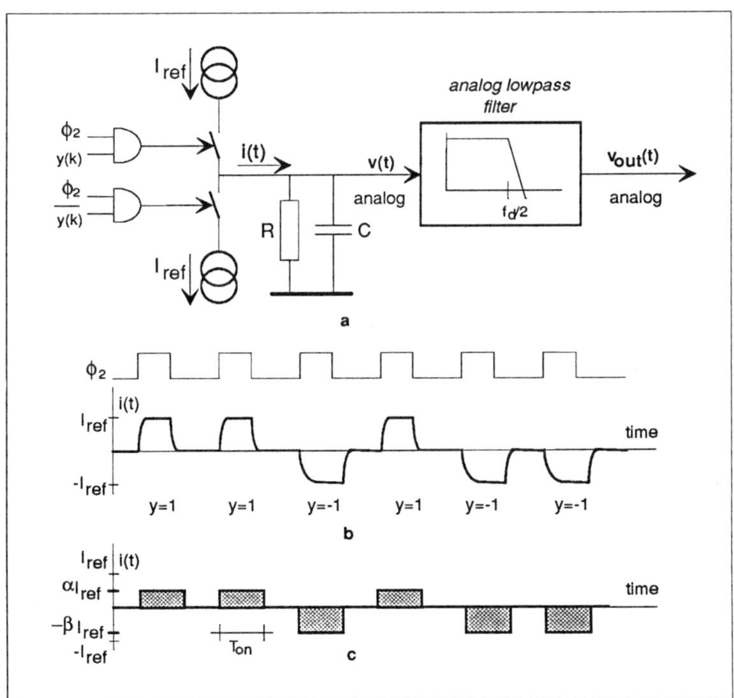

Fig. 5.8: Principle schematic of a current driven DAC

5.3.b. The signal degradation due to harmonic distortion

In the same way as for the voltage driven DAC, the harmonic distortion can be calculated from a nonlinear DC transfer characteristic. When a digital DC signal D is applied to the DAC input, the pulse density of y(k), p, is expressed by (5.17). The charge placed on the capacitor C by the current sources during the k-th clock period is given by (see fig. 5.8.c.):

$$Q(k) = [\alpha . I_{ref} - Y_1 . V] . T_{on} \quad \text{if } y(k) = 1 \tag{5.28.a}$$

$$= [-\beta . I_{ref} - Y2 . V] . T_{on} \quad \text{if } y(k) = -1 \tag{5.28.b}$$

In this expression, Y_1 and Y_2 are the current source output conductances. V is the DC value of v(t). T_{on} is the switch on-time. α and β are correction factors to take the finite switching speed into account. They are both close to one.

The long-term average charge stored on this capacitor has to equal zero. Hence,

$$p . [\alpha . I_{ref} - Y_1 . V] . T_{on} + (1-p) . [-\beta . I_{ref} - Y_2 . V] . T_{on} - V . T/R = 0 \tag{5.29}$$

where T is the clock period.

Combining this expression with (5.17) yields the nonlinear relation between D and V:

$$V = \frac{(\alpha-\beta) . I + (\alpha+\beta) . I . D}{[2/R . T/T_{on} + Y_1 + Y_2] + D . [Y_1 - Y_2]} \tag{5.30.a}$$

$$= \frac{(\alpha-\beta) . I_{ref}}{[2/R . T/T_{on} + Y_1 + Y_2]} + \frac{(\alpha+\beta) . I_{ref}}{[2/R . T/T_{on} + Y_1 + Y_2]} . D + \frac{Y_1 - Y_2}{[2/R . T/T_{on} + Y_1 + Y_2]^2} . D^2 + \ldots \tag{5.30.b}$$

The first term in this last expression is an offset voltage, the second term represents the linear gain and the third term is the second-order nonlinearity.

The second harmonic distortion for a quasi-static digital input signal can be calculated from expressions (5.30.b) and (2.9):

$$HD_2 = \frac{1}{2} . \frac{Y_1 - Y_2}{[2/R . T/T_{on} + Y_1 + Y_2]} . A \approx \frac{1}{4} . \frac{T_{on}}{T} . \frac{Y_1 - Y_2}{R} . A \tag{5.31}$$

where A is the digital signal input amplitude. α and β are approximated by one.

This expression is similar to expression (5.22). In the same way as for the voltage driven DAC, a mismatch between the current source output conductances causes second harmonic distortion [7]. However, with the use of cascode transistors, the current source output conductances can be made very small. A much better distortion performance can be obtained compared with a voltage driven DAC.

Expression (5.31) can be reformulated as

$$HD_2 \; = \; \frac{1}{4} \cdot [V_{early,1}^{-1} - V_{early,2}^{-1}] \cdot |V| \qquad\qquad (5.32.a)$$

$$= \frac{1}{4} \cdot \frac{V_{early,2} - V_{early,1}}{V_{early,2}} \cdot \frac{|V|}{V_{early,1}} \qquad\qquad (5.32.b.)$$

where $|V|$ is the output signal amplitude and V_{early} stands for the current source Early voltage [8]:

$$|V| \; = \; I_{ref} \cdot R \cdot \frac{T_{on}}{T} \cdot A \quad \text{and} \quad V_{early,i} \; = \; I_{ref}/Y_i \qquad\qquad (5.33)$$

E.g. for a signal amplitude of 1 Volt, for Early voltages in the order of 1000 Volt and with 20% mismatch between the Early voltages, formula (5.32) yields a second harmonic distortion of -86 dB. This is equivalent to a resolution of 14 bit.

5.3.c. The signal degradation due to clock jitter

The average current I(k) during the k-th clock period is depicted in Fig. 5.8.c. It approximately equals

$$I(k) \; = \; y(k) \cdot I_{ref} \cdot T_{on}/T \qquad\qquad (5.34)$$

Due to clock jitter, T_{on} can deviate from its nominal value, resulting in an error current:

$$\Delta I(k) \; = \; y(k) \cdot I_{ref} \cdot \Delta T_{on}/T \qquad\qquad (5.35)$$

These expressions are similar to equations (5.23) and (5.24). The in-band noise caused by this jitter is the same as for a voltage driven DAC. The SNR is expressed by (5.26).

5.3.d. An integrated example of a current driven DAC

In Fig. 5.9, a differential implementation of the current driven DAC of Fig. 5.8.a. is depicted, intended for ISDN applications [4]. The required signal bandwidth is 80 kHz.

During ϕ_1, both switches S_1 and S_2 are closed and the current from M_5 is equally divided over M_3 and M_4. When there are no mismatches in the current mirrors, the output currents $i_1(t)$ and $i_2(t)$ are zero.

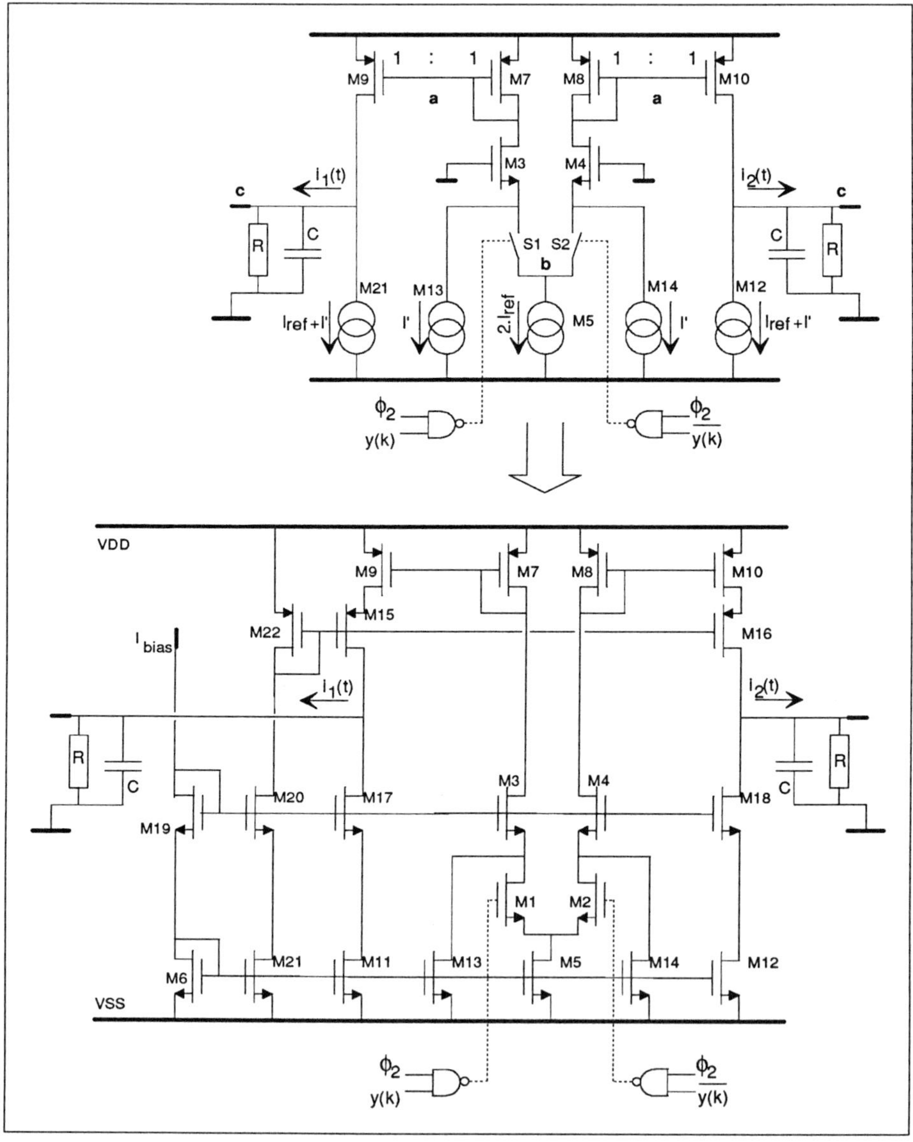

Fig. 5.9: schematic of the current driven DAC

During ϕ_2, one of the switches is open, depending on the value of y(k). For instance when y(k) is one, S_1 is open. All the current from M_5 flows into M_4. Hence, $i_2(t)$ equals I_{ref} while $i_1(t)$ is equal to -I_{ref}. For y(k) equal to minus one, the operation is similar but the signs of the output currents are opposite.

Without the current sources M_{13} and M_{14}, no current would flow through M_3, M_7, M_9 or M_{15} (M_4, M_8, M_{10} or M_{16}) when S_1 (S_2) is open. These transistors would cut off and a large time would be required to turn these transistors on again. Hence, the switching speed is improved by the current sources M_{13} and M_{14}. In Fig. 5.10, a SPICE simulation of the output current switching transients is shown. The decay time constant of the switching transients is determined by the internal poles on nodes a and b in Fig. 5.9. As can be seen, the currents are settled after about 4 nsec.

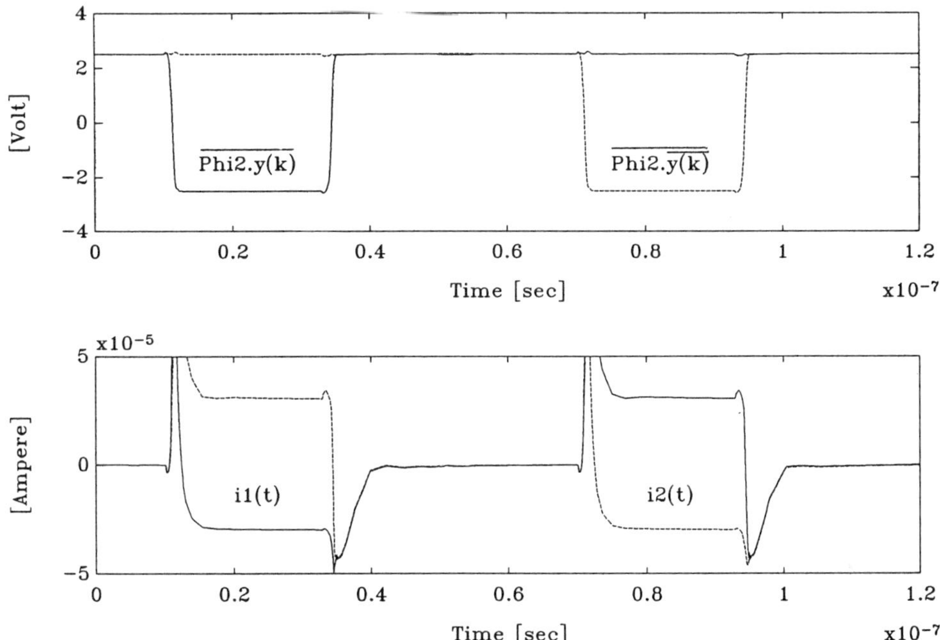

Fig. 5.10: Simulated output current waveforms

The average output current during the k-th clock period, at one circuit output, equals approximately:

$$I_2(k) \ = \ -I_1(k) \ = \ y(k) \cdot I_{ref} \cdot T_{on}/T \tag{5.36}$$

where T is the clock period and T_{on} is the switch on-time. For I_{ref} equal to 30 µA, T_{on} equal to half of the clock period and for resistors of 60 kOhm, the output voltage range is approximately between -1 Volt and +1 Volt.

The capacitors and resistors at the output nodes form passive lowpass filters. The cut-off frequency is 88 kHz. This signal band covers the ISDN band. Since polysilicon resistors and double-poly capacitors with a good linearity are used, the distortion due to these filters is much smaller than the distortion expressed by (5.32).

Except for the parasitic capacitances and output conductances at the output nodes, circuit nonlinearities cause no harmonic distortion since they only determine the shape of the transients in Fig. 5.10. As mentioned before, this causes no harmonic distortion since these shapes are identical for all the pulses with the same sign. Similarly, clock feedthrough causes no linearity problem either.

The major harmonic distortion is due to the current source output conductances and due to the nonlinear parasitic capacitances at the output nodes. The former one is expressed by (5.32). The parasitic junction capacitances on the nodes c in Fig. 5.9. are in parallel with the 30 pF filter capacitances. For parasitic capacitances in the order of 50 fF with an estimated nonlinearity of 30%/V, the harmonic distortion in each side of the circuit can be approximated as

$$HD_2 = \frac{1}{2} \cdot \frac{50 \ fF * 30\%/V}{30 \ pF} \cdot |V| \qquad (5.37)$$

where IVI is the output signal amplitude. For an output amplitude of 1 Volt, this yields -72 dB. With the differential approach in Fig. 5.9, the even harmonics are further suppressed and a distortion improvement of about 20 dB can be expected.

This circuit is realised in a 2 µm CMOS double poly, double metal technology. The transistors of the current driven DAC occupy 0.05 mm^2 of die area. The filter resistors and capacitors are 0.3 mm^2 large.

The performance of this DAC is tested with the measurement setup depicted in Fig. 5.11. (see also Fig. 3.45.c). A calculated PDM signal, obtained with the routine of section 3.7, is stored in a buffer RAM memory. For this calculation, a mathematical sinusoidal signal with a period of 1024 clock cycli and an amplitude of 0.6 times the full scale is applied to the modulator input. Fig. 5.12.a. shows the power spectrum of the calculated signal. The in-band quantisation noise power is 104 dB below the fundamental signal amplitude.

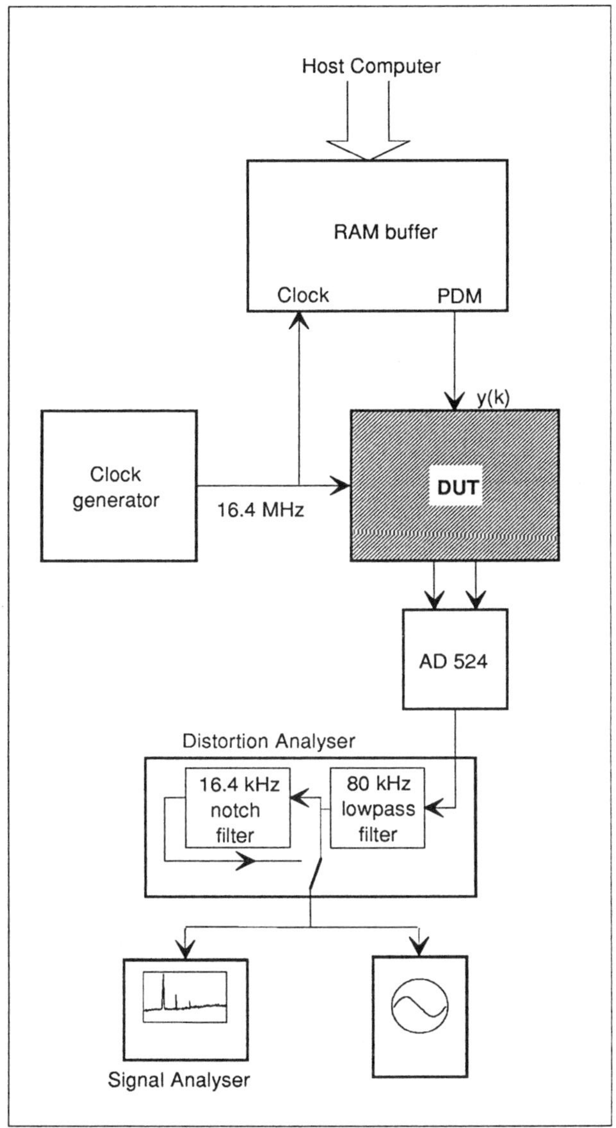

Fig. 5.11: The DAC measurement setup

A sequence of 1048576 (= 2^{20}) PDM bits is applied to the DAC PDM input. The differential DAC output signal is converted to a single-ended voltage with an instrumentation amplifier (e.g. an AD524 from Analog Devices).

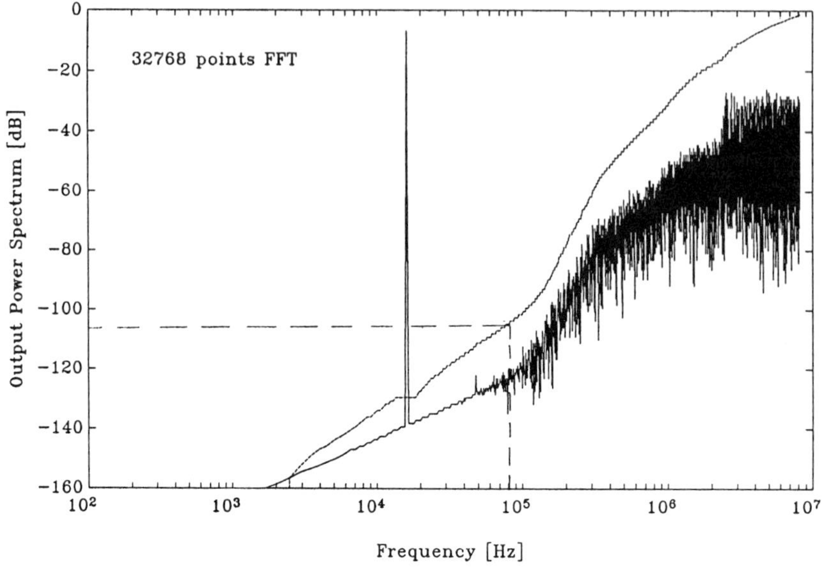

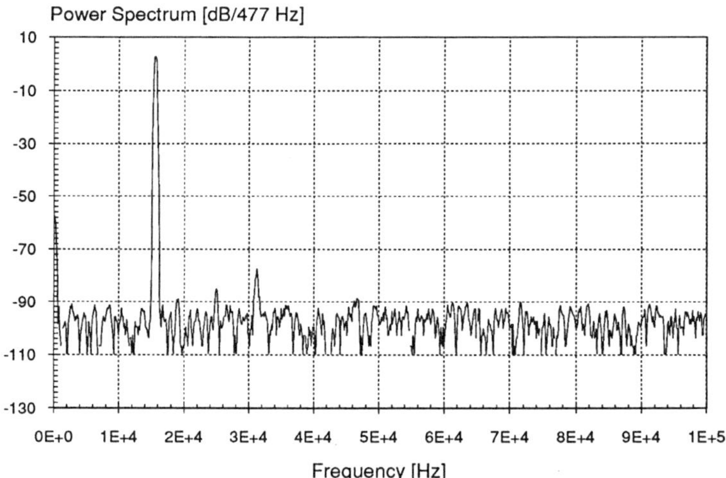

Fig. 5.12: a) Calculated input spectrum
b) measured output spectrum of the DAC

With a low-distortion 80 kHz lowpass filter from a distortion analyser (e.g. a HP339 from Hewlett Packard), the high-frequency quantisation noise is removed. The resulting power spectrum can be visualised with a Signal Analyser,

In Fig. 5.12.b, the DAC output spectrum, measured at the AD524 amplifier output, is depicted. The measured second harmonic is 79 dB below the fundamental signal. The measured SNR is 75 dB. This rather large noise is mainly due to clock jitter (see expression (5.26)). The total resolution is 12 bit.

In Fig. 5.13, the DAC output after filtering with the 80 kHz lowpass filter (see Fig. 5.11.) is depicted. The signal amplitude is about 1.2 Volt (0.6 times 2 Volt) and the signal frequency is 16 kHz (16.4 MHz/1024). The second curve is the remainder of the signal after suppressing the fundamental frequency with the notch filter (see Fig. 5.11). As can be seen, the distortion components are about 10000 times smaller than the fundamental signal.

The DAC characteristics are collected in Table 5.1.

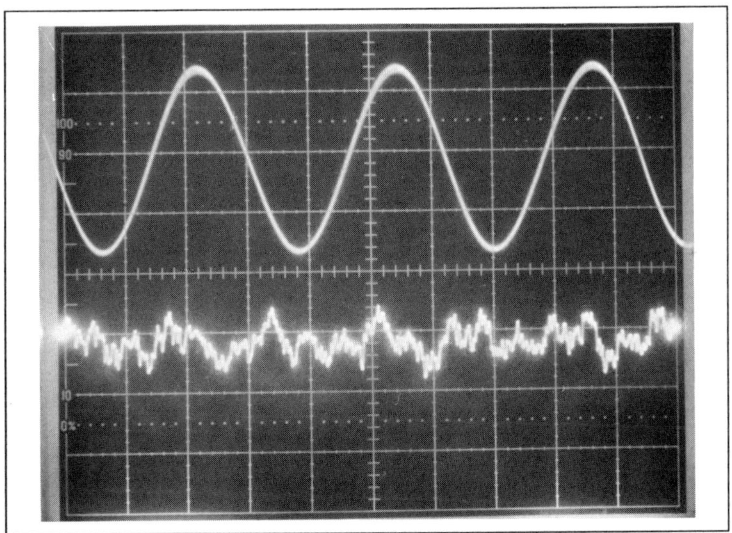

Fig. 5.13: The DAC output signal (see Fig. 5.11)
a) after the 80 kHz lowpass filter in Fig. 5.11: (0.5 V/div.)
b) after the 16 kHz notch filter: (200 µV/div.)

V_{DD}, V_{SS}	+/- 2.5 Volt
Supply current	255 µA
Clock Frequency	16.4 MHz
Signal Bandwidth	80 kHz
Output amplitude	2 V differential
SNR	75 dB
HD_2	-79 dB
Resolution	12 bit

Table 5.1: Characteristics of the Current driven DAC

5.3.e. Conclusions

With a current driven DAC, better distortion specifications can be obtained compared with a voltage driven DAC. However, the sensitivity for clock jitter is not improved.

A design example achieves a resolution of 12 bit for a signal bandwidth of 80 kHz. Although this resolution is not sufficient for some DSP applications that require a very high dynamic range [2], the circuit of Fig. 5.9. is interesting for other applications, for instance for a low-power programable analog waveform synthesizer. With a PDM pattern loaded from a digital memory, various analog waveforms can be generated [4].

5.4. A SWITCHED-CAPACITOR DAC

5.4.a. The signal transfer

In Fig. 5.14.a, a third possible solution is depicted for the analog reconstruction in a Sigma-Delta DAC [2] [9]. This circuit is a Switched-Capacitor lowpass filter.

During the phase ϕ_2 of each clock period, a charge is placed on the integration capacitor C_2, equal to:

$$Q(k) \; = \; -V_{ref} \cdot C_1 \qquad \text{if } y(k) = 1 \tag{5.38.a}$$

$$Q(k) \; = \; V_{ref} \cdot C_1 \qquad \text{if } y(k) = -1 \tag{5.38.b}$$

or, in a closed expression:

$$Q(k) \; = \; -V_{ref} \cdot C_1 \cdot y(k) \tag{5.39}$$

If we assume for a moment that the switch on-resistances are zero and that amplifier is ideal with an infinite GBW, infinite slew rate and infinite DC gain, the charge transient is infinitely fast and the amplifier input node $v_{in}(t)$ (see Fig. 5.14.a.) is always at ground potential. The charge current i(t) consists of a series of Dirac Impulses is as shown in Fig. 5.14.b. If the frequency spectrum of y(k) is denoted with $Y(j\omega)$, the charge current spectrum equals

$$I(j\omega) \; = \; -V_{ref} \cdot C_1 \cdot Y(j\omega) \tag{5.40}$$

The voltage over the parallel connection of R and C_2 can be calculated as:

$$V_R(j\omega) \; = \; - \; \frac{V_{ref} \cdot RC_1 \cdot Y(j\omega)}{1+j\omega \cdot RC_2} \tag{5.41}$$

$$= \; - \; \frac{V_{ref} \cdot RC_1 \cdot [D(j\omega)+E(j\omega)]}{1+j\omega \cdot RC_2} \tag{5.42}$$

where $D(j\omega)$ and $E(j\omega)$ are the Fourier transforms of the digital input d(k) and the quantisation noise e(k) respectively. These expressions illustrate that the PDM signal y(k) is converted to an analog voltage. Since the amplifier input node $v_{in}(t)$ is at ground voltage, the output voltage v(t) equals minus $v_R(t)$. Due to the R-C_2 filter and the analog lowpass filter, only the low-frequency content of this voltage appears at the output. The DAC output signal contains the requested analog output signal plus some in-band quantisation noise.

Compared with the circuits described in sections 5.2. and 5.3, the circuit of Fig. 5.14.a. has the advantage of a lower sensitivity for clock jitter. The charge placed on the integration capacitor C_2 during the k-th clock cycle, given by expression (5.39), is independent of the clock on-time. This is the major advantage over the current driven DAC of Fig. 5.9.

Again, this circuit is theoretically insensitive to component mismatches. A mismatch between the reference voltages or an error on the ratio C_1-C_2 causes only an offset voltage or a gain error, but no harmonic distortion. Similarly, the amplifier offset is not important.

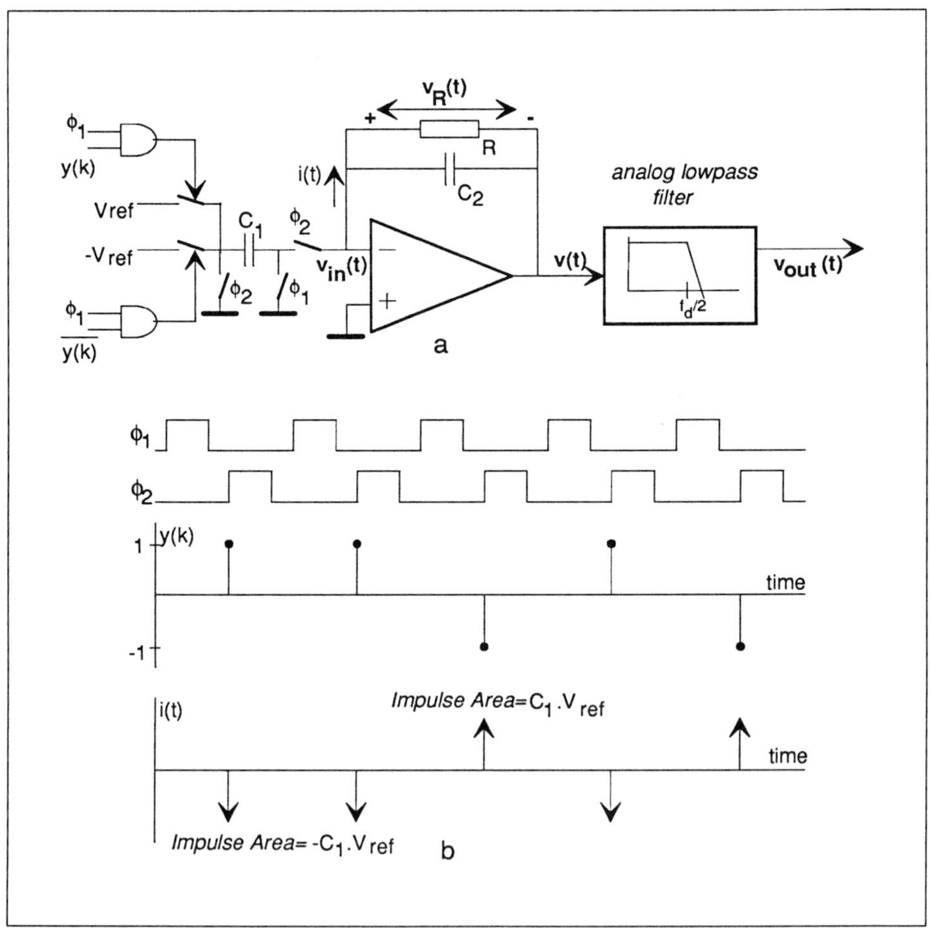

Fig. 5.14: A Switched-Capacitor DAC

5.4.b. The harmonic distortion

In practical circuits, the amplifier has a finite, nonlinear DC gain, a finite GBW and a maximum slew rate. In Fig. 5.15.b, the waveform $v_{in}(t)$ at the input of a real amplifier is depicted.

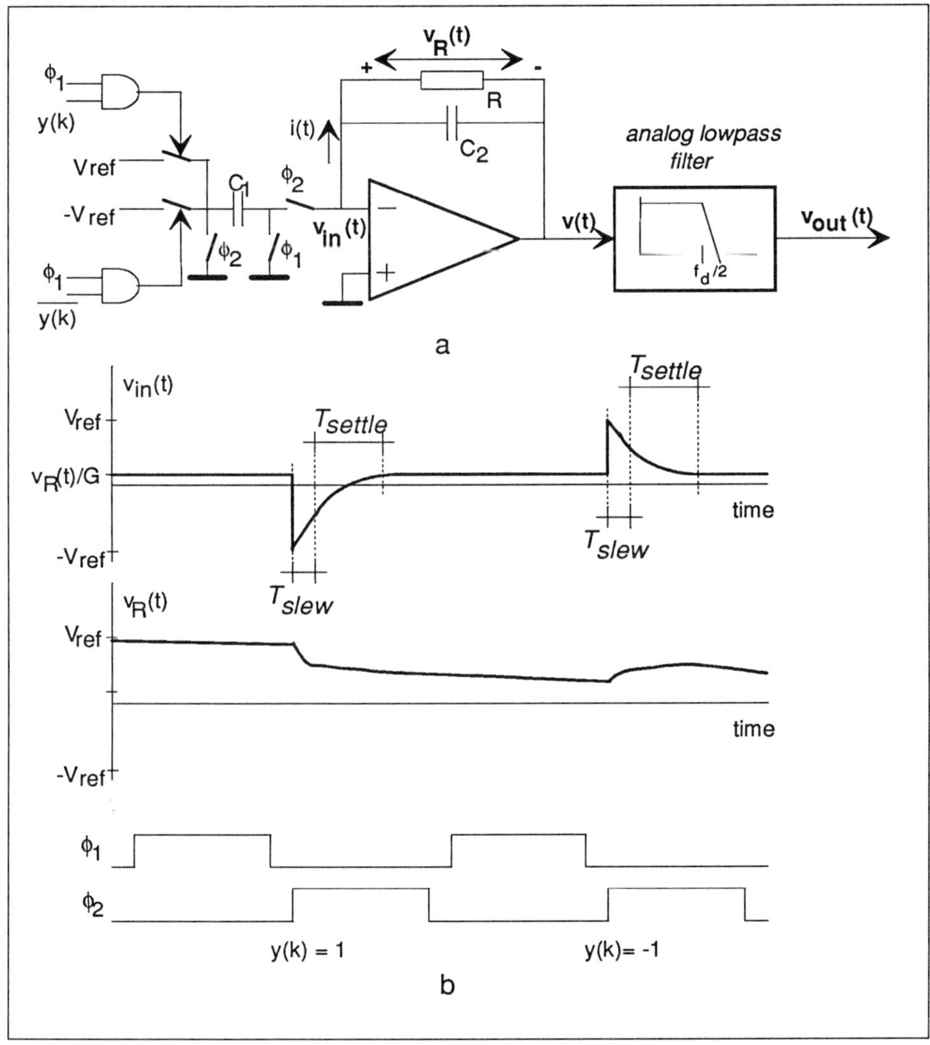

Fig. 5.15: Waveforms for a real Switched-Capacitor DAC

Due to the finite amplifier slew rate and settling time, it takes some time to transfer the charge from C_1 to C_2. The amplifier input node is not constantly at ground potential. The output voltage v(t) can be expressed as:

$$v(t) \;=\; v_{in}(t) \;-\; v_R(t) \tag{5.43}$$

The low-frequency contents of $v_{in}(t)$ and $v_R(t)$ appear at the amplifier output and pass the lowpass filter. When these are nonlinear functions of the DAC input signal, distortion is generated.

In the same way a for the two previous circuits, a nonlinear DC transfer characteristic can be calculated to obtain a harmonic distortion figure. However, this calculation requires a detailed analysis of the switching transients at the amplifier input, including the effects due to the amplifier nonlinear gain and slew rate [10] [11] [12]. We just mention here the results of these tedious calculations.

The amplifier nonlinear gain and the amplifier slew rate are the two major effects that cause harmonic distortion. When the amplifier nonlinear gain is characterised by the coefficients A_k, according to expression (2.6), the second harmonic distortion due to the nonlinear gain can be calculated as:

$$HD_2 \;=\; -\frac{1}{2} \cdot \frac{A_2(\omega_o)}{A_1(\omega_o)} \cdot \frac{1}{1+T(\omega_o)} \cdot \frac{|V|}{A_1(\omega_o)} \tag{5.44.a}$$

$$\text{with} \quad \omega_o \;=\; 1/RC_2 \tag{5.44.b}$$

In these expressions, $|V|$ is the amplifier output amplitude and T is the loop gain. This distortion term is nothing else than the closed loop distortion of the amplifier, measured at the R-C_2 lowpass filter frequency.

The second distortion term is due to the amplifier slew rate. In multiple stage amplifiers, the amplifier slew rate for an increasing output voltage can be different from the slew rate for a decreasing output voltage. This causes harmonic distortion:

$$HD_2 \;=\; \frac{1}{4G} \cdot \frac{T_{slew}}{T} \cdot \frac{SR_n - SR_p}{SR_p} \cdot A \tag{5.45}$$

In this expression, SR_n and SR_p are the amplifier slew rate for a decreasing output voltage and for an increasing voltage respectively, T is the clock period, T_{slew} is the slew time (see Fig. 5.15.b.), G is the amplifier gain at the R-C_2 filter frequency ($G=GBW.2\pi.RC_2$) and A is the digital input signal amplitude (between zero and one). Note that expression (5.45) is similar to expressions (5.22) and (5.31). Where a mismatch between switch on-resistances causes distortion in a voltage driven DAC and a mismatch between output conductances causes distortion in a current driven DAC, a mismatch between slew rates causes distortion in a Switched-Capacitor DAC.

5.4.c. Conclusions

The Switched-Capacitor DAC of Fig. 5.14. is superior over the continuous-time circuits described in sections 5.2. and 5.3, because of its lower sensitivity for clock jitter. This explains why the D-to-A converters with the highest dynamic range described in the literature are Switched-Capacitor circuits [2] [9]. Note that a similar conclusion is obtained in section 4.3.b. for a Sigma-Delta ADC. However, a low-noise, high-gain, highly linear amplifier with a fast settling time and with a large slew rate is required. The power consumption and the circuit complexity of this amplifier will be larger.

5.5. SUMMARY

This chapter is devoted to the analog circuits in a Sigma-Delta DAC. Although the principle operation of these circuits is very simple, the expected performance can be severely degraded by analog circuit nonidealities. Especially for the harmonic distortion, it is not clear à priori which circuit nonlinearities cause a performance degradation. With the calculation method for the nonlinear DC transfer characteristic, presented in this chapter, insight can be obtained and most of the common errors can be avoided.

Three circuits are compared for their harmonic distortion and sensitivity for clock jitter.

For the voltage driven DAC, both the harmonic distortion performance and the sensitivity for clock jitter are too large.

The current driven DAC allows a better linearity. A design example is discussed, achieving 75 dB SNR and -79 dB harmonic distortion. This yields a resolution of 12 bit for a signal bandwidth of 80 kHz.

With a Switched-Capacitor realisation, the sensitivity for clock jitter can be further improved. However, a complex amplifier with a fast settling time, a low-noise performance, a good linearity and a large slew rate is required. The power consumption will therefore be much larger.

5.6. REFERENCES

[1]   J. H. FISCHER: "Noise sources and calculation techniques for Switched-Capacitor Filters" - *IEEE J. Solid-State Circuits* vol. SC-17 No.4, Aug. 1982, pp. 742-752

[2]   P. J. NAUS *et al:* "A CMOS Stereo 16-bit D/A Converter for Digital Audio" - *IEEE J. Solid-State Circuits* vol. SC-22, No. 3, June 1987, pp. 390-395

[3]   Y. MATSUYA *et al*: "A 17-bit Oversampling D-to-A Conversion Technology Using Multistage Noise Shaping" - *IEEE J. Solid-State Circuits*, vol. SC-24, no. 4, August 1989, pp. 969-975

[4]   D. SALLAERTS *et al*: "A Single-Chip U-Interface Transceiver for ISDN" - *IEEE J. Solid-State Circuits* vol. SC-22, No. 6, Dec. 1987 pp. 1011-1021

[5]   private communication with ir. D. Rabaey, dr. ir. J. Sevenhans and ing. D. Haspeslagh, Alcatel Bell Telephone, Antwerp, Belgium

[6]   F. HEYRMAN, P. MEULEMANS, B. MAES: "12 bit D/A Convertor voor digitale telefonie" - Eindwerk KU Leuven, 1988

[7]   F. OP'T EYNDE, P. MEULEMANS, F. HEYRMAN, B. MAES and W. SANSEN: "A calulation method to predict in-band harmonic distortion of Σ–Δ D/A Converters" - *Proc. ISCAS* 1989, Portland, pp. 671-674

[8]   P. GRAY, R. G. MEYER: "Design and Analysis of Analog Integrated Circuits" - Wiley, New-York, 1977

[9]   L. R. CARLEY: "A Noise-Shaping Coder Topology for 15+ Bit Converters" - *IEEE J. Solid-State Circuits* vol. SC-24 no. 2, April 1989 pp. 267-273

[10]   K. LEE, R. MEYER: "Low-Distortion Switched-Capacitor Filter Design Techniques" - *IEEE J. Solid-State Circuits*, vol. SC-20 no. 6, Dec. 1985, pp. 1103-1113

[11]   K. HALONEN: "Low Power High-Performance Switched-Capacitor Circuits for Data-Acquisition Systems" - Ph. D thesis, KU Leuven, October 1987

[12]   W. SANSEN, H. QIUTING, K. HALONEN: "Transient Analysis of Charge Transfer in SC Filters - Gain Error and Distortion" - *IEEE J. Solid-State Circuits*, vol. SC-22 no. 2, April 1987, pp. 268-276

INDEX